Weather Forecasting

Edited by Muhammad Saifullah

Published in London, United Kingdom

IntechOpen

Supporting open minds since 2005

Weather Forecasting
http://dx.doi.org/10.5772/intechopen.78861
Edited by Muhammad Saifullah

Contributors
Cong Thanh, Tran Tan Tien, Dao Nguyen Quynh Hoa, Emmanuel P. Paul Agbo, Luis Alberiko Gil-Alana, Toju Esther Babalola, Philip Gbenro Oguntunde, Ayodele Ebenezer Ajayi, Francis Omowonuola Akinluyi, Muhammad Saifullah, Muhammad Waqas, Sarfraz Hashim, Mohsin Khan, Muhammad Adnan, Rana Muhammad Adnan, Shiyin Liu, Yasir Latif, Mudassar Iqbal, Sher Muhammad

First published in London, United Kingdom, 2021 by IntechOpen
IntechOpen is the global imprint of INTECHOPEN LIMITED, registered in England and Wales, registration number: 11086078, 5 Princes Gate Court, London, SW7 2QJ, United Kingdom
Printed in Croatia

British Library Cataloguing-in-Publication Data
A catalogue record for this book is available from the British Library

Additional hard and PDF copies can be obtained from orders@intechopen.com

Weather Forecasting
Edited by Muhammad Saifullah
p. cm.
Print ISBN 978-1-83968-053-3
Online ISBN 978-1-83968-054-0
eBook (PDF) ISBN 978-1-83968-055-7

For EU product safety concerns:
IN TECH d.o.o., Prolaz Marije Krucifikse Kozulić 3, 51000 Rijeka, Croatia, info@intechopen.com or visit our website at intechopen.com.

Meet the editor

Dr. Muhammad Saifullah has research experience in the field of agricultural water resources and climate change. He obtained a Ph.D. from Hohai University, China. He is the recipient of two researcher awards from the Ministry of Education, China, as well as a second-place award from the Yellow River Institute of Hydraulic Research. He has published peer-reviewed international research articles and investigated different international and national research projects. Dr. Saifullah worked as an assistant professor at the University of Agriculture Faisalabad and the University of Faisalabad, Pakistan. Then, he worked as a researcher at the Institute of International Rivers and Eco-security, Yunnan University, China. Currently, he is an assistant professor in the Department of Agricultural Engineering, Muhammad Nawaz Shareef University of Agriculture, Multan, Pakistan.

Contents

Preface

Weather forecasting is the process of using atmosphere, land, ocean, and meteorology data to project the weather of a given location. Forecasting mainly depends on precipitation, atmospheric pressure, relative humidity, sunshine hours, and cloud cover. Nowadays, it relies on model-based numerical calculations. However, human efforts to select the appropriate forecast model are also involved, necessitating that the meteorologist have knowledge of model performance and biases. The uncertainty in weather forecasting is due to the complicated nature of the atmosphere, complex calculations, and lack of understanding of the atmosphere and related processes. The use of statistical models, artificial intelligence techniques, machine learning models, the Kalman filter, and other advanced methods help meteorologists reduce errors in weather predictions.

Weather warnings are important because governments use them to protect life and property. In addition, predicting temperature and precipitation is important for agriculture. According to the United Nations' Sustainable Development Goals for 2030, weather forecasting is integral to meeting targets 2 and 13: zero hunger and climate action, respectively.

This book presents recent developments in scientific research on weather and climate in the context of currently ongoing processes in the extreme environments of Asian, African, and European regions. It provides in-depth case studies from Pakistan, the United States, Vietnam, Nigeria, and Africa. The global and interdisciplinary results of these studies help us to understand and address the grand challenges of weather as well as its impact on society.

Weather forecasting and climate change have to be carefully considered and monitored in order to educate for advancement in weather forecasting and to implement strategies of resilience, adaptation, and mitigation for meeting the SDGs.

Dr. Muhammad Saifullah
Assistant Professor,
Department of Agricultural Engineering,
MNS University of Agriculture Multan,
Multan, Punjab, Pakistan

Section 1

Weather Forecasting

Chapter 1

The Role of Statistical Methods and Tools for Weather Forecasting and Modeling

Emmanuel P. Agbo

Abstract

The need to understand the role of statistical methods for the forecasting of climatological parameters cannot be trivialized. This study gives an in depth review on the different variations of the Mann-Kendall (M-K) trend test and how they can be applied, regression techniques (Simple and Multiple), the Angstrom-Prescott model for solar radiation, etc. The study then goes ahead to apply some of them with data obtained from the Nigerian Meteorological Agency (NiMet), and applying tools like the python programming language and Wolfram Mathematica. Results show that the maximum ambient temperature for Calabar is increasing ($Z = 2.52$) significantly after the calculated p-value <0.05 (significant level). The seasonal M-K test was also applied for the dry and wet seasons and both were found to be increasing ($Z = 3.23$ and $Z = 4.04$ respectively) after their calculated p-values <0.05. The relationship between refractivity and other meteorological parameters relating to it was discerned using partial differential equations giving the gradient of each with refractivity; this was compared with results from the correlation matrix to show that the water vapor contents of the atmosphere contributes significantly to the variation of refractivity. Multiple linear regression has also been adopted to give an accurate model for the prediction of refractivity in the region after the residual error between the calculated refractivity and predicted refractivity was minimal.

Keywords: meteorology, forecasting, python programming, climate, Mann-Kendall, multiple linear regression

1. Introduction

The importance of statistical modeling and forecasting of time series data, etc., cannot be overemphasized. The benefits ranges from easy interpretability arising from visualization of results to the removal of the mysticism factor for the layman. The word 'forecasting' has to do with predicting the future based on data from the past and present. This is regularly done by the analysis of trends.

A routine example might be the estimation of temperature trends for some specified future date. Compared to forecasting, prediction can be seen as a term which is more general.

Forecasting methods have been applied in different areas ranging from climatology, finance, foreign exchange, etc. This has been applied in different regions of the world for better prediction and simulation. The key distinction in Information and Communication Technology (ICT) is the fact that with this technology, we can make predictions and simulations from previously obtained data. This is true and can be applied for every area while paying attention to the rules that govern them.

In this study we will be applying some statistical methods which can be adopted for the forecasting of climatic (weather) parameters in different regions of the world.

It is important to note that the predictability of the atmosphere is not perfect, this brings into context the fact that although statistical methods are necessary, results obtained are not totally accurate which is why room for errors (uncertainties) are given, albeit, a trend can be observed [1]. Statistical methods have been applied in the study of different regions for example, Daniel S. Wilks in [1] buttressed on the use of these methods on the analyses of different regions that do not necessarily have the same climatic condition. This brings into context the fact that laws are true irrespective of the region, i.e. neglecting all other factors that have little contribution to weather, the same methods can be applied in different regions to yield accurate results.

Analysis of trends can be useful in depicting and predicting the changing patterns and erraticism of some climatic parameters. This analysis gives a proper knowledge about the changing conditions of the climate and its effects, by the evaluation of meteorological parameters.

A data scientist using any tool or software for modeling and forecasting is particularly interested in the progression of these parameters (meteorological) as a function of time(t) $f(t)$. The designers of navigation or monitoring systems cannot trivialize the importance of forecasting as this is a very important part of their system. The spatial and temporal changes of atmospheric parameters calls for the adoption of this analysis to discern the effects of some meteorological parameters on some variables; for example, see [2].

A very popular software for any data scientist that is willing to understand the nitty-gritty of weather forecasting is Python Programming. This paper will explain in detail the setup processes for this to help the layman get started. A dataset of temperature trend in Calabar, Nigeria will be used at the end of this chapter to test the processes explained for better visualization.

The applicability of results from forecasting cannot be underestimated because this is great information for people that depend on weather conditions like farmers, surfers, and event planners, etc. The accurate prediction of atmospheric parameters can go a long way in positively affecting the financials of the informed, as money can be saved by avoiding unnecessary cost during trying times [3]. Natural disasters like Tsunami can be predicted with the correlation of meteorological parameters, harnessing information as explained previously and then incorporating this information through machine learning into the design of forecasting systems.

We delve deeper into a review of statistical methods like the M-K test and its different variations, the Angstrom-Prescott model for the estimation of solar radiation, linear regression techniques, with a deep look into multiple linear regression which will be applied in predicting refractivity after obtaining the coefficients of the variables. Results will be obtained and explained.

2. Review of statistical tests/methodology

With the shift going on in the world of technology, the implementation of some time series forecasting methods will be explained as well as their python implementation techniques. We often use forecasting models on time series data for the estimation of future trends of meteorological parameters.

2.1 Statistical test for trend (Mann-Kendall trend test)

One of the most important and widely applied test for trends involving time series is the Mann-Kendall trend test. It is mostly used for environmental and hydrological data. The test is non parametric and does not necessitate the data conforming to a particular distribution, similarly, the sensitivity of the test due to an inhomogeneous series resulting to abrupt breaks is very low [4]. The null hypothesis H_o which says that there is no monotonic trend in the series, is tested against the alternative hypothesis H_1 which says that there is a trend in the series. The test is applied to cases where a range of data $x_i$is in agreement with the equation below;

$$x_i = f(t_i) + \varepsilon_i \tag{1}$$

$f(t_i)$is a function of time and ε_i are the range residuals with zero mean.
The Mann-Kendall test statistic S is calculated using the formula

$$S = \sum_{k=1}^{n-1} \sum_{j=k=1}^{n} \text{sgn}\left(x_j - x_k\right) \tag{2}$$

where;

$$\text{sgn}\left(x_j - x_k\right) = \begin{cases} +1; & if\left(x_j - x_k\right) > 0 \\ 0; & if\left(x_j - x_k\right) = 0 \\ -1; & if\left(x_j - x_k\right) < 0 \end{cases} \tag{3}$$

n in Eq. (2) is the number of data values in the studied series. The advantage of this test is that it can handle the situation where data values are incomplete with respect to the number of years or months, etc. [4]

In the case where n is greater than or equal to 10 (10 and above), we adopt the normal approximation (Z).

To find the variance of S, '$VAR(S)$', we compute Eq. (4) below.

$$VAR(S) = \frac{1}{18}\left[n(n-1)(2n+5) - \sum_{p=1}^{g} t_p\left(t_p - 1\right)\left(2t_p + 5\right)\right] \tag{4}$$

From the equation, the number of data values is represented by n, the number of equal of tied groups is represented by g, and the number of data values in the pth group is represented by t_p.

We now use the results from $VAR(S)$ to find the test statistic Z

$$Z = \begin{cases} \dfrac{S-1}{\sqrt{VAR(S)}}; & S > 0 \\ 0; & S = 0 \\ \dfrac{S+1}{\sqrt{VAR(S)}}; & S < 0 \end{cases} \tag{5}$$

A decreasing trend can be discerned from results of Eq. (5) when the value of Z is negative and an increasing trend when Z is positive (**Table 1**).

The significance of an increasing or decreasing trend is observed when the p-value of the series is lower than the significance level ($\propto$), in this case, we can say

Significance level (∝)	Required n
0.1 (10%)	≥ 4
0.05 (5%)	≥ 5
0.01 (1%)	≥ 6
0.001 (10%)	≥ 7

Table 1.
Significance level (∝) required for given numbers of data.

there is a trend observed trend in the series [5]. The adoption of different significant levels with respect to the number of given data values n is given in **Table 1**.

The classification of this probability/significance level is important because results can be confused to be entirely true. We need to understand that the significance level of say 0.05, means that there is a 5% probability that a mistake will be made while rejecting the null hypothesis H_o. Similarly, a significance level of 0.01 means that there is a 1% probability that a mistake will be made while rejecting H_o.

2.2 Regression analysis

The two easiest ways to forecast time series data by observation are the simple regression and the moving average, they both depend on historical data. The former demands mere observation of the previous trend and drawing up an extrapolation from there; this can be somewhat less accurate. The moving average has been used for forecasting meteorological data like rainfall (See reference [6]). Analyzing with regression has to do with the relationship one variable which is dependent has with one or more independent variables. We use them to check for models showing the strength of relationship between the variables and any possible future relationships [1].

2.2.1 Simple linear regression

This regression variation is based on the assumption that the two variables (dependent and independent variable) show a linear relationship between the intercept and the slope, similarly, there is no residual error in this regression and the value is constant across all observations.

$$Y = \pm mX \pm c + e \tag{6}$$

Y is the dependent variable.
X is the independent variable.
m is the value of the slope.
c is the intercept.
e is the residual error.
The regression is depicted by a straight line describing the Eq. (6) above (**Figure 1**).

2.2.2 Multiple linear regression

This model is similar to that of simple linear regression, but the only exception is that it has multiple independent variables, unlike that of simple linear regression which has just the one. This can be represented by Eq. (7);

$$Y = \pm m_1X_1 \pm m_2X_2 \pm m_3X_3 \pm c + e \tag{7}$$

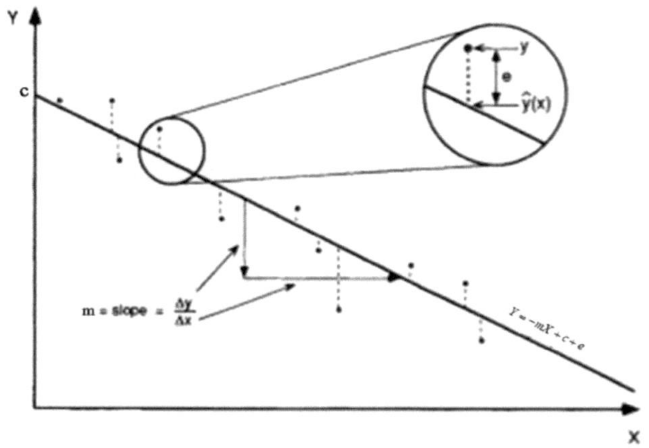

Figure 1.
Schematic illustration of simple linear regression. The regression line, $Y = -mX + c + e$, is chosen as the one minimizing some measure of the vertical differences (the residuals) between the points and the line. The residual e is the difference between the data point and the regression line.

Y is the dependent variable.
X_1, X_2, X_3 are the independent variables.
m_1, m_2, m_3 are the values of the slopes.
c is the intercept.
e is the residual error.

One thing to note about multiple linear regression is that the independent variables must not be collinear, i.e., they do not have to have a high correlation coefficient between each other, else there will be difficulty in assessing the relationship between the dependent and independent variables.

We also need to take note that before multiple linear regression is performed on range of data values, a linear relationship must exist between each independent variable and the dependent variable. The amount of residual error must be almost constant at each point in the model. The multiple linear regression will be applied to study and predict refractivity trend in Calabar, Nigeria. This was done with the 'statsmodel' package in python programming and results have been displayed in section 2.5.

A perfect meteorological equation that this regression technique can be applied to is the refractivity equation recommended by the International Telecommunication Union (ITU) shown in Eq. (8);

$$N = 77.6\frac{P}{T} + 3.73 \times 10^5 \frac{e}{T^2} (N - units) \tag{8}$$

P is the Atmospheric Pressure (hPa).
e is the Atmospheric Vapor Pressure (hPa).
T is the Absolute Temperature (K).

Eq. (8) shows the relationship between refractivity (dependent variable) and meteorological parameters (ambient temperature, atmospheric pressure, and vapor pressure) which are all independent variables.

This has been applied in [7] modeling the meteorological parameters for the accurate determination of refractivity. These meteorological parameters (Ambient Temperature, Atmospheric Pressure and Relative Humidity) have been obtained from the Nigeria meteorological Agency (NiMet), Calabar.

Results have been presented in section 2.5. From Eq. (8), we obtain the atmospheric vapor pressure e from the relation;

$$e = \frac{e_s H}{100}(hPa) \tag{9}$$

e_s is the saturated vapor pressure (hPa) calculated from;

$$e_s = 6.11\exp\left(\frac{17.26(T - 273.16)}{T - 35.87}\right)(hPa) \tag{10}$$

2.3 Review of the application of simple linear regression analysis in climatology (the Angstrom-Prescott model)

The linear regression technique can be applied to find the relationships between an independent variable and the dependent variable. We can see the explanation of this from Eq. (6).

One major example of the benefits of linear regression is the estimation of the Angstrom-Prescott coefficients of the Angstrom-Prescott model for a particular region as this relates to solar radiation. The Angstrom-Prescott model is given by [8];

$$\frac{H}{H_0} = a + b\frac{n}{N} \tag{11}$$

where the monthly average daily extraterrestrial radiation is given by H_0, H is the monthly average daily global radiation in Wh/m^2/day. n is the actual sunshine duration in a day for a particular region (hours), N is the monthly mean length of the day in hours. The Angstrom-Prescott empirical coefficients are given by a and b. The linear regression technique has been adopted by Srivastava and Pandey [8] to find by a and b. Comparing Eq. (6) to Eq. (9) we have that;

$$\begin{aligned} &\frac{H}{H_0} = Y\,(\text{variable}) \\ &\frac{n}{N} = X\,(\text{variable}) \\ &b = m = slope \\ &a = c = Y\,\text{intercept} \end{aligned} \tag{12}$$

This shows that if we have the variables '$\frac{H}{H_0}$ and $\frac{n}{N}$', we can get the values of a and b, from our Y intercept and slope respectively. Getting these constant values for specific regions will help us forecast future trends.

For better understanding, the extraterrestrial radiation H_0 is given by the equation [9];

$$\begin{aligned} H_0 = &\frac{24 \times 3600 \times I_{SC}}{\pi} \times \left[1 + 0.33\cos\left(\frac{360 \times d}{365}\right)\right] \\ &\times \left[\cos\phi\cos\delta\sin\omega + \frac{\pi\omega}{180}\sin\phi\sin\delta\right] \end{aligned} \tag{13}$$

Here, I_{SC} is the solar constant with a value of 1367 W/m^2, d represents the day of the year (from January 1st to December 31st); taking January 1st as 1 and December 31st as 365 or 366 (in the case of a leap year). The latitude of the study location, the declination angle and the sunset hour angle are represented by ϕ, δ, and ω respectively. $\omega = \cos^{-1}(-\tan\phi\tan\delta)$. The declination angle can be obtained from [9].

$$\delta = 23.45\sin\left[360\left(\frac{284 + d}{365}\right)\right] \tag{14}$$

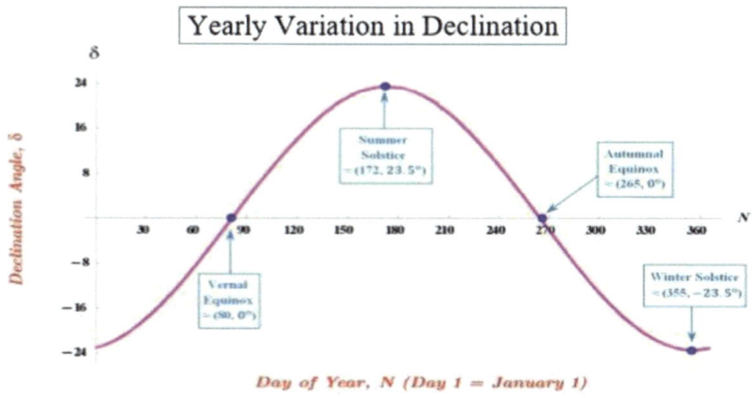

Figure 2.
Yearly variation of declination angle δ with respect to the days of the year.

The monthly mean length of the day (in hours) can be obtained from [9].

$$N = \frac{2\omega}{15} \tag{15}$$

The above equations can be applied to estimate the coefficients using linear regression. By this we can use these coefficients to predict solar radiation for a given region.

We know that the declination angle ranges from $-23.5 \leq \delta \leq +23.5$. From **Figure 2**, we can see that the declination angle is 0 ° C at the Verbal and Autumnal Equinox, while the angles are −23.5 and + 23.5 at the summer and winter solstice respectively. It is easy to see why this has a huge effect on the variation of Global solar radiation.

Klein in 1977 [10] recommended average days of the various months and corresponding angle of declination as in **Table 2**.

2.4 Calculus in climatology

Applying calculus in environmental science is important in predicting a lot of things. It can be applied to understand the impacts of parameters on the variations of other parameters that they relate to. It is important to know that calculus is the 'mathematical study continuous change' so this can be applied in climatology to discern the impacts of some parameters on the "continuous change" of others [11–13].

Writing the refractivity equation in terms of relative humidity H, by substituting (10) into (9), and the into (8), we have;

$$N = 77.6\frac{P}{T} + 3.73 \times 10^5 \frac{6.11 \exp\left(\frac{17.26(T-273.16)}{T-35.87}\right) \times 0.01H}{T^2} (N - units) \tag{16}$$

Similarly, obtaining refractivity in terms of the saturated vapor pressure e_s using Eq. (8) and (9) gives;

$$N = 77.6\frac{P}{T} + 3.73 \times 10^5 \frac{e_s H}{100T^2} (N - units) \tag{17}$$

Now applying partial differentials to the equations for refractivity; Eqs. (8), (16), and (17), we obtain partial differentials relating each parameter to refractivity;

$$\frac{\partial N}{\partial P} = \frac{77.6}{T}$$

$$\frac{\partial N}{\partial T} = -\left(77.6\frac{P}{T^2} + 7.46 \times 10^5 \frac{e}{T^3}\right)$$

$$\frac{\partial N}{\partial H} = \frac{22790.3 \exp\left[\frac{17.26(-273.16+T)}{-35.87+T}\right]}{T^2} \qquad (18)$$

$$\frac{\partial N}{\partial e} = \frac{3.73 \times 10^5}{T^2}$$

$$\frac{\partial N}{\partial e_s} = \frac{3.73 \times 10^3 \times H}{T^2}$$

From monthly Temperature, Humidity and Atmospheric pressure data obtained for 2005–2018 from the archives of the Nigerian meteorological agency (NiMet) Calabar, the atmospheric vapor pressure and the saturated vapor pressure can be obtained by applying these parameters in Eqs. (9) and (10) (**Figure 3**).

2.5 Python implementation for Mann-Kendall trend test

With the python software installed, the next step will be installing an IDE (integrated development environment). The easiest IDE to use is the Jupyter Notebook. This IDE displays results as you code.

We will walk you through the processes for analyzing data by using the data for Calabar in the south of Nigeria, collected from the archives of the Nigeria meteorological agency (NiMet). Research has been done in this area in climatology [14–18], but with the application of python and the Mann-Kendall test can give more meaning to time series data.

Month	Date	Day of the year (d)	declination angle (δ)
January	17	17	−20.9
February	16	47	−13
March	16	75	−2.4
April	15	105	9.4
May	15	135	18.8
June	11	162	23.1
July	17	198	21.2
August	16	228	13.5
September	15	258	2.2
October	15	288	9.6
November	14	318	−18.9
December	10	344	−23

Table 2.
Recommended average days for various months and their corresponding declination angles [10].

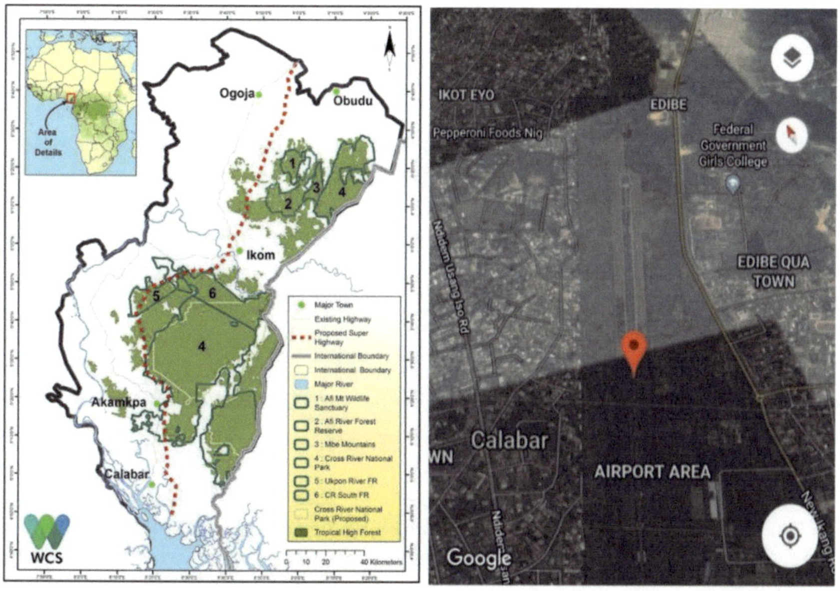

Figure 3.
Map of study area showing Calabar as a coastal area (left) and the exact location of the Nigerian meteorological agency (NiMet) where the data was obtained (right).

We need to install the python package for the Mann-Kendall test called '*pymannkendall*'. To install this package, the following python packages are required;

- Numpy
- Scipy

For handling and cleaning data we need the '*pandas*' package, and for data visualization we need the '*matp*lotlib' package.

We want to analyze maximum ambient temperature data for 20 years in Calabar.

In the Jupyter notebook, the first step will be to import the respective packages. We must also note that for our examples in the Appendices, we stored the excel file containing the data used for the analysis in the same folder as the python file for easy reference.

Appendix A shows the process of importing the installed packages required for the analysis into the workspace.

Before we perform the Mann-Kendall test, we need to import the excel file titled '*Temperature*' in which the table is stored, in a sheet name called '*MAX*'. See *Appendix B.*

Appendix C shows how the Mann-Kendall original test is performed after importing the packages and data. We assigned the name of the imported data file as '*Max*' and set the significance level (∝) to the default 5% (0.05); this can be adjusted by the user to his/her preference. Results were obtained and displayed in *Appendix C.*

We now perform the seasonal M-K test for the dry season variation, we import the excel file titled '*Temperature*', the date column will be an index column. The sheet name of the excel file in which the data is stored is called 'dry'. This implementation can be seen from *Appendix D.*

Appendix E shows the seasonal M-K test python implementation for the dry season variation. By setting the significance level (∝) to the default 5% (0.05), and the period to 4, which stands for the 4 months of the dry season in the study area (November to February), we have satisfied the criteria for the seasonal M-K test.

For the wet season variation, the excel file titled '*Temperature*' will be imported and the date will be an index column. The sheet name is called 'wet'. *Appendix F* shows the implementation code for this importation.

We can now perform the seasonal Mann-Kendall test on the wet season data. *Appendix G* shows this. The Seasonal Mann-Kendall test of the imported file we assigned the name 'wet' has been achieved by setting the significance level (∝) to the default 5% (0.05); this can be adjusted by the user to his preference. We also set the period to 8, which stands for the 8 months of the wet season in the study area (March to October).

There are other variations of the Mann-Kendall test along with their python implementation [19]. These can be used depending on the data obtained and the aim of the test.

1. Hamed and Rao Modified MK Test (*hamed_rao_modification_test*): This test addresses serial correlation issues

2. Yue and Wang Modified MK Test (*yue_wang_modification_test*): This is also a variance correction method for considered serial autocorrelation proposed by Yue, S., & Wang, C. Y. (2004). User can also set their desired significant n lags for the calculation.

3. Modified MK test using Pre-Whitening method (*pre_whitening_modification_test*): This test pre-whitens the time series before applying the trend test

4. Modified MK test using Trend Free Pre-Whitening method (*trend_free_pre_whitening_modification_test*): This test removes the trend component from the series before pre-whitening and the applying the trend test

5. Multivariate MK Test (*multivariate_test*): As the name implies, this test is for multivariate (multiple) parameters. This can be used for monthly data, where each month can be considered as a parameter.

6. Regional MK Test (*regional_test*): As the name implies, this calculates the trend at a regional scale

7. Correlated Multivariate MK Test (*correlated_multivariate_test*): Unlike the Multivariate MK test, this test is also a multivariate mk test, but the parameters are correlated.

8. Correlated Seasonal MK Test (*correlated_seasonal_test*): This test is similar to the seasonal MK test, but in this is used when the time series is significantly correlated with previous seasons/months

9. Partial MK Test (*partial_test*): Due to the fact that in some studies, many factors can affect the dependent parameters, so we overcome this by inputting one dependent parameter and an independent parameter.

10. Theil-Sen's Slope Estimator (*sens_slope*): This test method proposed by Theil (1950) and Sen (1968) [20] is applied to estimate the magnitude of the monotonic trend.

11. Seasonal Theil-Sen's Slope Estimator (*seasonal_sens_slope*): This test method considers the seasonal effect of the Theil-Sen's Slope Estimator.

3. Results and discussion

3.1 Results

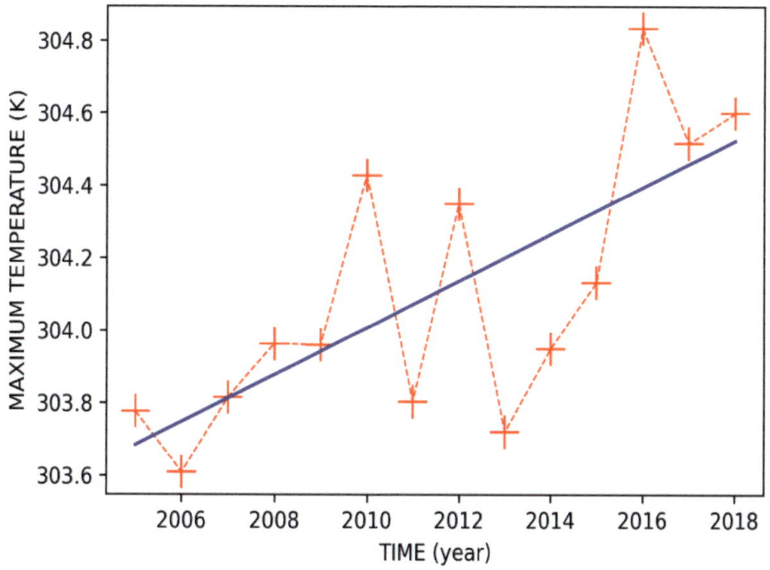

Figure 4.
Mann-Kendall trend of maximum ambient temperature.

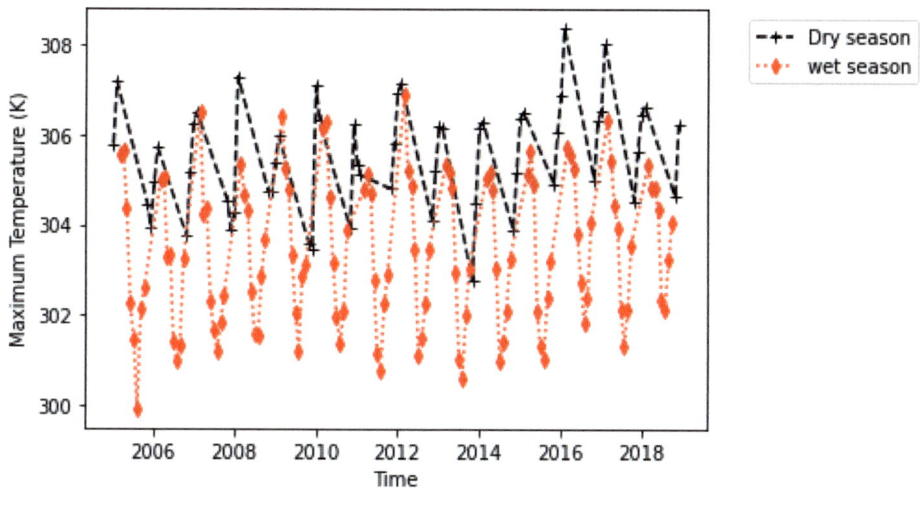

Figure 5.
Seasonal trend of maximum ambient temperature for dry and wet season.

3.2 Discussion

For the annual variation in **Figure 4**, results show that there is a trend in the series as the p-value is less than the significance level (0.05). The positive Z value (observed from *Appendix C*) shows that the series is increasing. We can conclude that the maximum ambient temperature variation is increasing, and it is doing so with significance, the slope of the trend can be observed from the results in *Appendix C.*

For the dry season variation observed in **Figure 5**, results show that there is a trend in the series. The positive Z value of the dry season trend observed from *Appendix E* shows that the series is increasing. We can conclude that the maximum temperature variation in the dry season is increasing significantly as the calculated p-value is less than the significance level (0.05), the slope of the trend can be observed from results in *Appendix E.*

For the wet season variation observed also in **Figure 5**, results show that there is a trend in the series. The positive Z value from *Appendix G* shows that the series is increasing. We can conclude that the maximum temperature variation in the dry season is increasing significantly as the calculated p-value is less than the significance level (0.05), the slope of the trend can be observed from the results in *Appendix G.*

These results are in agreement with Agbo et al. [2] for the same region.

3.2.1 Relationship between refractivity and meteorological parameters

To understand the relationship between refractivity and all parameters relating to it, we adopt Eq. (18) by substituting obtained and calculated data.

From the data obtained at the Nigerian Meteorological Agency (NiMet) Calabar, and adopting Eq. (9) and (10) we obtain the total annual values for the meteorological parameters as;

P = 1005.97 hPa; H = 85.71%; T = 300.28 K; e = 30.71 hPa; e_s = 35.94 hPa. Substituting these values into the equations in Eq. (18), we obtain;

$$\begin{aligned}
\frac{\partial N}{\partial P} &= 0.258425 \\
\frac{\partial N}{\partial T} &= -0.0196183 \\
\frac{\partial N}{\partial H} &= 1.48436 \\
\frac{\partial N}{\partial e} &= 4.13672 \\
\frac{\partial N}{\partial e_s} &= 3.62832
\end{aligned} \tag{19}$$

Results from the gradients of the differential equations in Eq. (19) show that the vapor pressure and saturated vapor pressure contributes more to the variation of refractivity. The relative humidity similarly has a high gradient; this can be physically explained by relating the water vapor content of the atmosphere to the variation of refractivity.

The correlation plot of refractivity and all other meteorological parameters is shown in **Figure 6**. Results agree with that of the differential equations in Eq. (19). As seen in Eq. (19), the correlation plot showed that the atmospheric vapor pressure and relative humidity had high positive relationships with refractivity. The saturated vapor pressure however has a low correlation coefficient compared to the high gradient in Eq. (19); this can be interpreted thus; that the variation of the saturated vapor pressure has a relatively high contribution to the variation of refractivity, but the saturated vapor pressure does not have a similar trend to that of refractivity.

3.2.2 Application of multiple linear regression in climatology

Multiple linear regression has been applied to relate refractivity with obtained meteorological parameters. The goal is to obtain an equation that relates refractivity

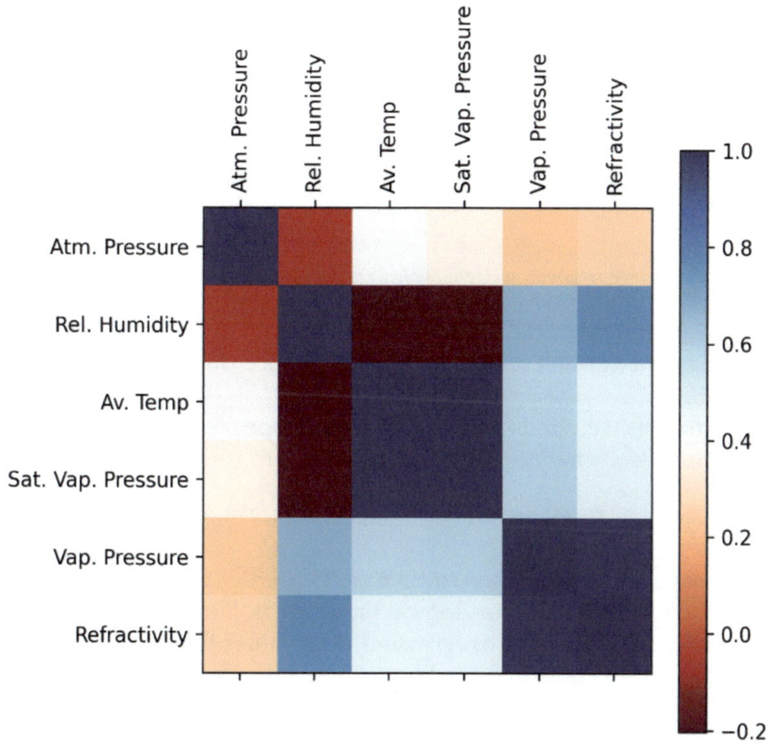

Figure 6.
Correlation matrix of atmospheric parameters and refractivity.

to meteorological parameters through Multiple Linear Regression (MLG). Using Eq. (8) to calculate refractivity, we show results in **Table 3**. As part of the conditions for carrying out multiple linear regression, we have to test for collinearity

Year	Pressure	Temperature	Humidity	Refractivity (N)
2005	1005.15	300.23	87.15	388.88
2006	1005.38	300.17	85.38	385.73
2007	1005.50	300.10	84.84	384.74
2008	1005.44	300.21	86.00	387.11
2009	1005.83	300.29	83.36	383.75
2010	1005.46	300.71	83.26	385.62
2011	1005.80	300.12	87.49	388.83
2012	1005.75	300.44	87.97	391.57
2013	1005.74	300.03	87.07	387.69
2014	1005.92	299.79	85.15	383.38
2015	1006.55	300.03	85.68	385.83
2016	1006.97	300.70	85.69	389.73
2017	1007.09	300.58	86.33	389.90
2018	1007.02	300.52	84.58	386.84

Table 3.
Data of obtained meteorological parameters and refractivity.

	C	S_e	*t Stat*	*P-value*	*Lower 95%*	*Upper 95%*
Intercept	1617.97	51.05	−31.69	2.30×10^{-11}	−1731.72	−1504.23
Pressure	0.17	0.05	3.25	8.75×10^{-03}	0.05	0.28
Temperature	5.68	0.13	44.90	7.22×10^{-13}	5.39	5.96
Humidity	1.53	0.02	68.62	1.05×10^{-14}	1.48	1.58

Table 4.
Output of the multiple linear regression showing the coefficients (C) of each parameter and their standard error (S_e).

between the independent variables. We see from the correlation matrix in **Figure 6** that the independent variables are not collinear, hence this satisfies the criteria for carrying out MLG.

From our analysis we obtain the coefficients (slopes) of the variables (meteorological parameters) and the intercept from **Table 4** to form the equation below;

$$Refractivity(N) = 1.53\,Humidity + 0.17\,Pressure + 5.68\,Temperature - 1617.97 \quad (20)$$

The above equation can be used to accurately predict the variation of refractivity, given the values of the meteorological parameters. **Table 4** shows these results obtained from the multiple linear regression. The values for the predicted refractivity (Predicted N) was gotten from Eq. (20) by substituting the values of the meteorological parameters. This equation is more straight forward that the equation recommended by ITU as all the variables and coefficients are all linear with respect to refractivity.

Figure 7 shows the trend of refractivity calculated from Eq. (8) with that of predicted refractivity, calculated from Eq. (20). The residual error seen from **Table 5** shows relatively constant values (in agreement with our MLG conditions), and a small deviation from the original values of refractivity.

From **Table 4** probability values (p-values) of the parameters are all less than the significance level (5% = 0.05; 95% confidence level), this shows that the variation agrees with the alternative hypothesis and shows a trend relating the independent variables to the dependent variables.

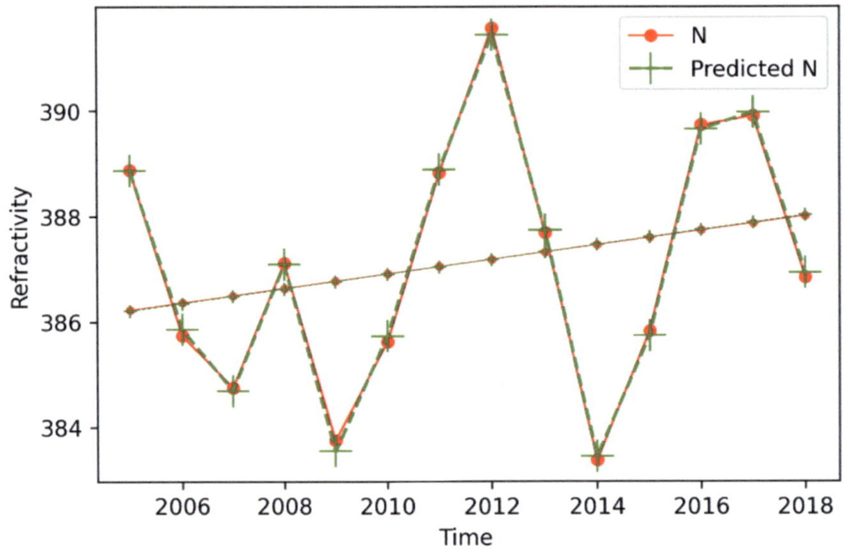

Figure 7.
Comparison plot of annual refractivity and predicted refractivity.

Year	N	Predicted N	Residuals
2005	388.88	388.87	0.005
2006	385.73	385.85	−0.121
2007	384.74	384.69	0.056
2008	387.11	387.09	0.025
2009	383.75	383.54	0.204
2010	385.62	385.72	−0.109
2011	388.83	388.89	−0.060
2012	391.57	391.45	0.124
2013	387.69	387.74	−0.048
2014	383.38	383.45	−0.074
2015	385.83	385.74	0.091
2016	389.73	389.65	0.076
2017	389.90	389.97	−0.072
2018	386.84	386.93	−0.095

Table 5.
Residual output derived from the results of the coefficients, showing the predicted refractivity values compared to the refractivity values to give the residuals.

Results from **Figure 7** show the minimal error between the predicted refractivity and the calculated refractivity. **Table 5** shows the values for both as well as the residual error between them. This shows that the error is small and thus, Eq. (20) can be adopted for the prediction of refractivity for the study area. This equation can be modified and refractivity N can be gotten in terms of other parameters like the saturated vapor pressure and the atmospheric vapor pressure.

4. Conclusion

There are myriads of ways in which weather can be forecasted and this arises from the understanding of basic meteorological parameters and how they behave in the atmosphere; and also from the understanding of the role of statistics in climate research [21]. Research in this area has been reviewed to give a better understanding of the different techniques for analyzing trends; which include, Linear Regression (Multiple and Simple), the Mann-Kendall trend test [22, 23] (to test for trends in a time series variation), the Angstrom-Prescott model for estimating solar radiation as well as the python implementation of some various techniques.

The multiple linear regression technique was applied to model an equation to accurately predict the trend for refractivity in the study location, the simple linear regression technique has been explained as well as accurate methods for its application in the predicting/estimation of the Angstrom-Prescott coefficients. These coefficients can be gotten for specific regions and can be accurately applied to predict solar radiation in that region.

Results from the multiple linear regression gave an accurate model for the prediction of refractivity in the region after the residual error between the calculated refractivity and predicted refractivity was minimal.

The Mann-Kendall original and seasonal test has been applied to analyze the maximum temperature in Calabar, Nigeria for the annual and seasonal (dry and wet season) variation respectively, and results show that the annual, dry season and wet season had increasing variations (after having positive Kendall Z-values of 2.52, 3.23, 4.04 respectively) and they were all increasing significantly at 5% (0.05) level of significance after their p-values were all less than 0.05 agreeing with Agbo and Ekpo [23].

The relationship between refractivity and other meteorological parameters relating to it was discerned using partial differential equations giving the gradient of each with refractivity; this was compared with results from the correlation matrix to show that the water vapor contents of the atmosphere contributes significantly to the variation of refractivity.

Acknowledgements

The author will like to acknowledge the Nigerian Meteorological Agency (NiMet) Calabar for providing the necessary data for applying in this study.

The author will also like to express his thanks and appreciation to the editor, whose comments greatly improved the chapter.

Conflict of interest

The author declares no conflict of interest.

A. Appendices

A.1 Appendix A

Input:

```
import numpy as np
import pandas as pd
import pymannkendall as mk
import matplotlib.pyplot as plt
%matplotlib inline
from pandas import ExcelWriter
from pandas import ExcelFile
from matplotlib.figure import Figure
```

This will import the above installed packages into the workspace.

A.2 Appendix B

```
Max = pd.read_excel("Temperature.xlsx", 'MAX' index_col= 'YEAR')
```

The excel file titled '*Temperature*' will be imported and the data will be an index column. The sheet name is called '*MAX*'.

We can now perform the Mann-Kendall test

A.3 Appendix C

Input

```
mk.original_test(Max, alpha=0.05)
```

Output

```
Mann_Kendall_Test(trend='increasing', h=True,
p=0.011793457077065028, z=2.518264946676251,
Tau=0.5164835164835165, s=47.0, var_s=333.6666666666667,
slope=0.06763844012453053, intercept=303.5218288324476)
```

A.4 Appendix D

```
dry=pd.read_excel("Temperature data.xlsx", 'Sheet2', index_col= 'YEAR')
```

The excel file titled '*Temperature*' will be imported and the data will be an index column. The sheet name is called 'dry'.

We can now perform the Mann-Kendall test

A.5 Appendix E

Input

```
mk.seasonal_test(dry, alpha=0.05, period=4)
```

Output

```
Seasonal_Mann_Kendall_Test(trend='increasing', h=True,
p=0.001232892414896325, z=3.231159219618304,
Tau=0.3269230769230769, s=119.0, var_s=1333.6666666666667,
slope=0.08467049808428379, intercept=305.1036046113848)
```

A.6 Appendix F

```
wet=pd.read_excel("Temperature data.xlsx", 'Sheet3', index_col=
'YEAR')
```

The excel file titled '*Temperature*' will be imported and the data will be an index column. The sheet name is called 'wet'.

We can now perform the Mann-Kendall test

A.7 Appendix G

Input

```
mk.seasonal_test(wet, alpha=0.05, period=8).
```

Output

```
Seasonal_Mann_Kendall_Test(trend='increasing', h=True,
p=5.126153098378161e-05, z=4.049799512953561,
Tau=0.28846153846153844, s=210.0, var_s=2663.3333333333335,
slope=0.05741935483871145, intercept=302.85004032258064)
```

Author details

Emmanuel P. Agbo[1,2,3]

1 Cross River University of Technology, Calabar, Cross River State, Nigeria

2 Lafarge Africa PLC, Mfamosing Plant, Akamkpa, Cross River State, Nigeria

3 Osun State University, Osogbo, Osun State, Nigeria

*Address all correspondence to: emmanuelpaulagbo@gmail.com

References

[1] Statistical weather forecasting. In. Daniel S. Wilks S, editor. Elsevier: International Geophysics; 1995. pp. 159-232. DOI: 10.1016/S0074-6142(06)80042-2

[2] Agbo EP, Ettah EB, Eno EE. The impacts of meteorological parameters on the seasonal, monthly, and annual variation of radio refractivity. Indian Journal of Physics. 2021; 95: 195-207. DOI: 10.1007/s12648-020-01711-9

[3] Datta A, Si S, Biswas S. Complete Statistical Analysis to Weather Forecasting. In: Das A., Nayak J., Naik B., Pati S., Pelusi D. (eds) Computational Intelligence in Pattern Recognition. Advances in Intelligent Systems and Computing, 2020. vol. 999. Springer. Singapore. DOI: 10.1007/978-981-13-9042-5_65

[4] Alhaji UU, Yusuf AS, Edet CO, Oche CO, Agbo EP. Trend analysis of temperature in Gombe state using Mann Kendall Trend test. Journal of Scientific Research and Reports. 2018;**20**(1):1-9. DOI: 10.9734/JSRR/2018/42029

[5] Agbo EP, Ekpo CM, Edet CO. Analysis of the effects of meteorological parameters on radio refractivity, equivalent potential temperature and field strength via Mann-Kendall test. Theoretical and Applied Climatology. 2021 Feb; 143 (3): 1437-1456. DOI: 10.1007/s00704-020-03464-1

[6] Akrami SA, El-Shafie A, Naseri M, Santo CAG. Rainfall data analyzing using moving average (MA) model and wavelet multi-resolution intelligent model for noise evaluation to improve the forecasting accuracy. Neural Comput & Applic. 2014;**25**:1853-1861. DOI: 10.1007/s00521-014-1675-0

[7] Ajileye OO, Popoola OS, Kayode FF, Rabiu AB. Meteorological models for determination of surface radio refractivity over Nigeria. Current Journal of Applied Science and Technology. 2019;**36**(3):1-15. DOI: 10.9734/CJAST/2019/v36i330235

[8] Srivastava RC, Pandey H. Estimating Angstrom-Prescott coefficients for India and developing a correlation between sunshine hours and global solar radiation for India. International Scholarly Research Notices. 2013;**2013**

[9] Razmjoo A, Heibati SM, Ghadimi M, Qolipour M, Nasab JR. Using Angstrom-Prescott (A-P) Method for Estimating Monthly Global Solar Radiation in Kashan. J Fundam Renewable Energy Appl. 2016: 6(5). DOI: 10.4172/2090-4541.1000214

[10] Klein SA. Calculation of monthly average insolation on tilted surfaces. Solar Energy. 1977;**19**(4):325-329

[11] Ustrnul Z, Czekierda D. Application of GIS for the development of climatological air temperature maps: an example from Poland. Meteorological Applications: A journal of forecasting, practical applications, training techniques and modelling. 2005 Mar;**12**(1):43-50

[12] Testud J, Chong M. Three-dimensional wind field analysis from dual-Doppler radar data. Part I: Filtering, interpolating and differentiating the raw data. Journal of Applied Meteorology and Climatology. 1983 Jul 1;**22**(7):1204-1215

[13] Błażejczyk KR, Matzarakis AN. Assessment of bioclimatic differentiation of Poland based on the human heat balance. Geographia Polonica. 2007;**80**(1):63-82

[14] Agbo EP, Ekpo CM. Trend Analysis of the Variations of Ambient Temperature Using Mann-Kendall Test

and Sen's Estimate in Calabar. Southern Nigeria.

[15] Edet CO, Eno EE, Ettah EB, Kamgba FA. Seasonal Variation of Radio Refractivity in Calabar, Nigeria. measurements. 2017 Jun;6(6).

[16] Edet CO, Eno EE, Ettah EB. Monthly Variation of Radio Refractivity in Calabar, Nigeria. International Journal of Scientific Research Engineering & Technology. IJSRET: ISSN; 2017. pp. 2278-0882

[17] Edet CO, Eno EE, Ettah EB. Effects of Variations in Meteorological Parameters of Atmospheric Pressure. Relative Humidity and Temperature on Radio Refractivity in Calabar.

[18] Kamgba FA, Edet CO, Njok AO. Effects of some meteorological parameters on wind energy potential in Calabar, Nigeria. Asian journal of physical and chemical. sciences. 2017 Nov 16:1-7

[19] Hussain M. Mahmud I. pyMannKendall: a python package for non-parametric Mann Kendall family of trend tests. Journal of Open Source Software. 2019;**4**(39):1556. DOI: 10.21105/joss.01556

[20] Sen PK. Estimates of the regression coefficient based on Kendall's tau. Journal of the American statistical association. 1968 Dec 1;**63**(324): 1379-1389

[21] Zwiers FW, Storch HV. On the role of statistics in climate research. International Journal of Climatology. 2004;**24**:665-680. DOI: 10.1002/joc.1027

[22] Agbo EP, Ekpo CM, Edet CO. Trend Analysis of Meteorological Parameters, Tropospheric Refractivity, Equivalent Potential Temperature for a Pseudoadiabatic Process and Field Strength Variability. Using Mann Kendall Trend Test and Sens Estimate. arXiv preprint arXiv. 2020 Oct 9;**2010**: 04575

[23] Agbo EP, Ekpo CM. Trend Analysis of the Variations of Ambient Temperature Using Mann-Kendall Test and Sen's Estimate in Calabar, Southern Nigeria. InJournal of Physics: Conference Series 2021 Jan (Vol. 1734, No. 1, p. 012016). IOP Publishing. . DOI: 10.1088/1742-6596/1734/1/012016

Chapter 2

Evaluating the Performance of Different Artificial Intelligence Techniques for Forecasting: Rainfall and Runoff Prospective

Muhammad Waqas, Muhammad Saifullah, Sarfraz Hashim, Mohsin Khan and Sher Muhammad

Abstract

The forecasting plays key role for the water resources planning. Most suitable technique is Artificial intelligence techniques (AITs) for different parameters of weather forecasting and generated runoff. The study compared AITs (RBF-SVM and M5 model tree) to understand the rainfall runoff process in Jhelum River Basin, Pakistan. The rainfall and runoff of Jhelum river used from 1981 to 2012. The Different rainfall and runoff dataset combinations were used to train and test AITs. The data record for the period 1981–2001 used for training and then testing. After training and testing, modeled runoff and observed data was evaluated using R^2, NRMSE, COE and MSE. During the training, the dataset C2 and C3 were found to be 0.71 for both datasets using M5 model. Similar results were found for dataset of C3 using *RBF-SVM.* Over all, C3 and C7 were performed best among all the dataset. The M5 model tree was performed better than other applied techniques. GEP has also exhibited good results to understand rainfall runoff process. The RBF-SVM performed less accurate as compare to other applied techniques. Flow duration curve (FDCs) were used to compare the modeled and observed dataset of Jhelum River basin. For High flow and medium high flows, GEP exhibited well. M5 model tree displayed the better results for medium low and low percentile flows. RBF-SVM exhibited better for low percentile flows. GEP were found the accurate and highly efficient DDM among the AITs applied techniques. This study will help understand the complex rainfall runoff process, which is stochastic process. Weather forecasting play key role in water resources management and planning.

Keywords: Forecasting, Jhelum River, GEP, flow duration curve, RBF-SVM

1. Introduction

A long scientific challenge is weather forecasting. Accurate weather forecasting has a direct social and economic impact on the community [1]. Recently, Artificial Neural Networks are using for weather forecasting. The crucial parameter for weather forecasting is rainfall, which also generates runoff in watersheds area. This process is one of the fundamental factors in weather forecasting. The different approaches exist from physically, conceptual, modeling and artificial intelligence techniques (AITs) [2].

The rainfall-runoff process plays a vital role in sustainable water resources management. Pakistan economy depends on Agriculture. Water resources are crucial for agriculture, and most of the population livelihood depends on agriculture. Water storage is necessary, and the urban population's rapid growth [3, 4]. The efficient and precise modeling of the rainfall-runoff process is crucial in planning water resources management [5]. Urban water management, runoff forecasting, weather forecasting and irrigation system is become the current challenge due to the uncertainty of weather forecasting. Rainfall and geographical characteristics have importance to forecasting accurately rainfall-runoff process. Rainfall-runoff considers the diverse process and AIT used to transform rainfall into runoff [6]. Similar, the transformation of precipitation into runoff investigated in the science of hydrology by different researchers [7, 8], and runoff is a complex process [9]. During the forecasting mechanism of runoff, it becomes an essential issue in hydrology and water resources management.

Rainfall and other metrological parameter play a crucial role during the forecasting of weather, which is essential for runoff generation [10]. The rainfall-runoff process is non-linear. Simple AITs cannot model this non-linear process due to several hydrological variables such as evaporation, infiltration, rainfall intensity, watershed characteristics, and surface and groundwater interaction. During the last few decades, Artificial Neural Network (ANN), genetic programming (GP), Support vector machines (SVMs), Decision Trees (DTs), and adoptive Neuro-Fuzzy Inferences System (ANFIS) are considered most efficient in hydrology and water resources. Several researchers applied AITs to forecast rainfall-runoff [11–16]. American Society of Civil Engineering task committee applied ANNs in hydrology [17, 18]. ANNs and various algorithms were applied in a different region of the world [6, 19–22].

Many studies revealed that ANNs have some limitations and drawbacks in order to predict streamflow. These include stopping criteria, over fitting issue, low learning speed, back propagation problem, and some human intervention like learning epochs and learning rate [23]. Thus, there is a need to develop some approaches to overcome these problems and generate better results as compared with ANNs.

After 2000, Support vector machines SVMs, a new kernel-based approach, become famous and got advantages over ANN. In this study, SVM and DTs were used for rainfall-runoff modeling. Firstly, SVM was first developed after inspired by statistical machine learning theories (SMLTs) for complex problems like classification and regression [24, 25] emphasized the obstacles in rainfall-runoff prediction to recognize the best model and its relevant parameters. The modified form of SVM is the least square support vector machine (LS-SVM) which decrease the computational problem [26, 27]. In many types of research, SVM is used for different forecasting scenarios [28–30]. In this regard, several researchers applied the SVM. [31] publicized that in rainfall-runoff forecasting using past daily dataset using SVM and ANNs. The SVM found most efficient technique than ANN. [32] used the SVM technique using monthly time scale data for statistical downscaling of rainfall intensity. SVM model was successfully engaged and predicted daily rainfall [33]. Another DDM is [34] M5 model tree, and M5 model tree is DDM technique which uses divide and conquers method to split the dataset into subsets, which enable the system to distribute the multi-dimensional variables and automatically build a model on the inclusive quality benchmarks [34, 35] used SVM with RBF kernel function and polynomial functions to model the suspended sediment load of a basin Iran, which exposed that SVM with RBF function gives the most accurate modeling. In recent years, different hydrological components predicted by many researchers using M5 model tree such as; sedimentation transportation and estimation [36], rainfall-runoff prediction [37], prediction of flood events [38], monthly pan evaporation prediction [39], Modeling oblique load-carrying capacity [40] and Modeling algal a typical proliferation [41].

As mentioned above several ATIs were engaged for rainfall-runoff process forecasting but still there are some techniques which have not yet been evaluated such as RBF-SVM and the model tree M5. Himalayan rivers especially Jhelum River basin initiating primarily from >4000 masl, withstand tremendous amount of inhabitants downstream. Though, Jhelum River basin is very data limited, and hydrological data for hydro-meteorological factors is accessible mainly from the areas below 2000 masl. Since the high level of anthropological need on these rivers, it is essential to progress strategies and tactics based on the hydrology of these rivers [42–45]. Therefore, these AITs will be very necessary for forecasting of hydrological parameters especially rainfall-runoff processes. These AITs are actually need of this region where data management and acquiring of hydrological data is adamant.

Keeping the previous studies on modeling of rainfall-runoff processes in mind, this study was arranged in such a way for different employee AITs to achieve the primary objectives of this research as 1) to calibrate and validate the AITs (GEP, BRF-SVM and M5 model tree) for the modeling of the rainfall-runoff process; 2) to evaluate the best input combination for the applied AITs. To achieve these objectives, hydrological data of rainfall and runoff were employed to model this process. To evaluate models performances, some statistical evaluation parameters, i.e. determination coefficient (R2), coefficient of efficiency (COE), mean squared error (MSE), and normalized root mean square error (NRMSE), were used.

The input selection process for data-driven rainfall-runoff models is critical because input vectors determine the structure of the model and, hence, can influence model results. This chapter is arranged as follows. Section 1 "Introduction and Review literature" where all previously employed and selected methodology is discussed. Section 2, "Rainfall-Runoff forecasting", includes study area and data acquisition, which elaborates a brief summary description of the study area and dataset comprising nine gauges and runoff on past thirty years daily rainfall data dataset and Model fitness criterion, Trend analysis tests. Section 3," Methodology", summaries proposed AITs (RBF-SVM and M5 model tree). Section 4, "Results and Discussions", describes the analysis results of outputs of different applied AITs for modeling rainfall-runoff process and trend analysis of rainfall in different rainy seasons. Section 5, "Conclusion", accomplishes the study.

2. Materials and methods

2.1 Study area

The geographical Jhelum River basin situated at 33.14°N and 73.64°E. The drainage area of the basin is 33,867 km^2. It originates from Pir Panjale from the North-Western Part of the great Himalayan range and gets significant contributions to the flow from its tributaries. Kunhar and Neelum River fall in Jhelum River at Muzaffarabad. Poonch and Kanshi join the Jhelum at Mangla reservoir [46]. It is the Trans Boundary River between Pakistan and India. 56% of the area of the rivers occurred in India [47]. Jhelum River basin lays 25% under maximum snow accumulation. The dataset for the basin was collected from the Surface Water Hydrology Project (SWHP) from 1981 to 2012. It is mainly affected by monsoon rainfall. During the summer season, rain shadow of the Himalayas range makes Eastern Himalayan chronicles [48–50]. The rainfall station and flow station are shown in **Figure 1**. Western disturbance starts from December, and the moon soon starts

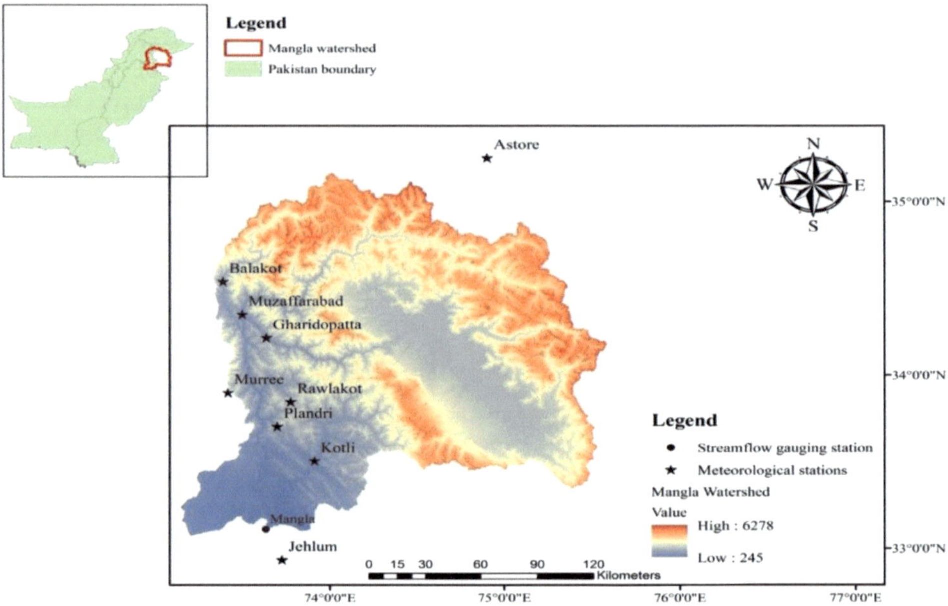

Figure 1.
Study area.

from June to September in every year [51]. The rainfall decreases from the northern part to the eastern region. The annual rainfall found to vary from 70—135% [52, 53]. Astor station also considered as previous researchers used for analysis [54].

2.2 Auto-correlation function (ACF) and partial auto-correlation function (PACF)

For the selection of proper input combinations of rainfall and runoff, the autocorrelation function (ACF) [55] and Cross-correlation function (CCF) [56] were employed for runoff data and rainfall-runoff data, respectively, with a 95% confidence level. From the **Tables 1** and **2**, it can be seen that the cross-relation in the rainfall and runoff dataset is poor, which may be an issue for modeling of rainfall-runoff phenomenon [57]. So, the partial autocorrelation was used between these two input variables. It is concluded from the results shown in **Table 1** that three lag times of rainfall and runoff datasets will be efficient for the modeling process. Based on results, the following input combinations were used in this study;

1. C1 Q(t-1)
2. C2 Q(t-2)
3. C3 Q(t-3)
4. C4 Q(t-2), Q(t-1), P(t-1), Pt
5. C5 Q(t-2), Q(t-1), P(t-2), P(t-1), Pt
6. C6 Q(t-3), Q(t-2), Q(t-1), Pt

Partial Auto-Correlation Function (PACF)								
Input Variable	**Coefficients**	**Standard Error**	**t Stat**	**P-value**	**Lower 95%**	**Upper 95%**	**Lower 95.0%**	**Upper 95.0%**
Q_{obs}	89224.45	30235.64	2.95	0.00	29953.86	148495.05	29953.86	148495.05
$P_{(t)}$	−1449.13	3534.18	−0.41	0.68	−8377.14	5478.88	−8377.14	5478.88
$P_{(t-1)}$	−840.46	3838.46	−0.22	0.83	−8364.94	6684.02	−8364.94	6684.02
$P_{(t-2)}$	−239.51	3838.46	−0.06	0.95	−7764.01	7284.99	−7764.01	7284.99
$P_{(t-3)}$	−1139.05	3534.17	−0.32	0.75	−8067.04	5788.94	−8067.04	5788.94

Table 1.
Partial auto-correlation function (PACF) between rainfall and runoff data.

Auto-correlation Function of Runoff Data			
Input Combinations	**Training Data**	**Testing Data**	**Whole Data**
Q_{obs}, Q_{t-1}	0.6785	0.9148	0.6785
Q_{obs}, Q_{t-2}	0.3194	0.8721	0.3194
Q_{obs}, Q_{t-3}	0.0008	0.8529	0.0008

Table 2.
Auto-correlation between runoff and rainfall data.

7. C7 Q(t-3), Q(t-2), Q(t-1), P(t-3), P(t-2), P(t-1), Pt

Where Q is discharge (m^3/sec), P is precipitation (mm), and it is Time (sec). There we created different time lags of Q and P to test and train the models, i.e. (t), (t-1), (t-2) and (t-3). These parameters are arranged to create different input combinations C1, C2, C3, C4, C5, C6 and C7, which are used for testing and training AI techniques to get better results.

2.3 Support vector machine (SVM)

A brief description of the SVM has been mentioned in this study, whereas the theory SVM [24] was discussed by many researchers in detailed, i.e. [28–30]. According to [24] in the SVM technique, independent variable x helps estimate the dependent variable y. The relationship between x and y was determined by the given function like other regression scenarios;

$$f(x) = (f(x).wT.\text{Ø}(x) + b) \tag{1}$$

$$f(y) = f(x) + \text{noise}) \tag{2}$$

where Ø is kernel function which can be defined as; it takes to input information and changes it into the desired shape. Various SVM algorithms practice diverse sorts of kernel functions. There are many kinds of these functions. i.e. sigmoid, polynomial, non-linear, linear, and RBF. b is a constant, w is the coefficient of vector, w and b are the constraints of the regression function. In contrast, noise is elaborated by error tolerance (e). During the training of the SVM model, a process of association of successive optimization of the error function in which can be achieved. There are two kinds of SVM models based on the error function, such as e-SVM (Regression I) and t-SVM (Regression II) [58]. In this study, BRF Regression, I is engaged because for prediction like rainfall-runoff purposes. [59, 60] proposed that the training time of SVM decreased by selecting the automatic RBF kernel function because it efficiently selected the proper kernel function constraints. As compared to V-fold validation is consumed less Time and more efficient. Let consider (x^{i}_{j}) j = 1… …. Ni Rd. is the dataset of i, and Ni is the number of training samples of i class. Whereas i = 1,2,3… …L and L is the number of classes in the dataset, then RBF is;

$$K(x, x', \sigma) = \exp.\left(-\frac{IIx - x'II^{1}/2}{2\,\sigma^{1}/2}\right) \tag{3}$$

K is a kernel function, (x', σ) are elements of R^{d} and σ element of R-0 which is corresponding constraints. It has two major possessions, i. the cosine value of training dataset ≥1, and it must be more than 0. ii. The norm in the dataset must be 1 [61] shown in **Figure 2**.

As in this study, RBF based kernel is used, so the following expression is used to calculate the mean of values;

$$b(\sigma) = \frac{\mathbf{1}}{\sum_{i=1}^{L}\sum_{i=1,J\neq 1}^{L} \boldsymbol{NiNj}} \sum_{i=1}^{L} \sum_{i=1,J\neq 1}^{L} \sum_{l=1}^{L} \sum_{k=1}^{L} \boldsymbol{k}(\boldsymbol{xl}\ (\boldsymbol{i}), \boldsymbol{xk}(\boldsymbol{j}), \boldsymbol{\sigma}) \tag{4}$$

Therefore, b(σ) is calculated in a pattern that (σ) is must be greater than 0 but not less or 0. The σ can be calculated in SVM based on RBF kernel function by solving the given steps;

1. To determine the best constraint, the given expression is optimized.

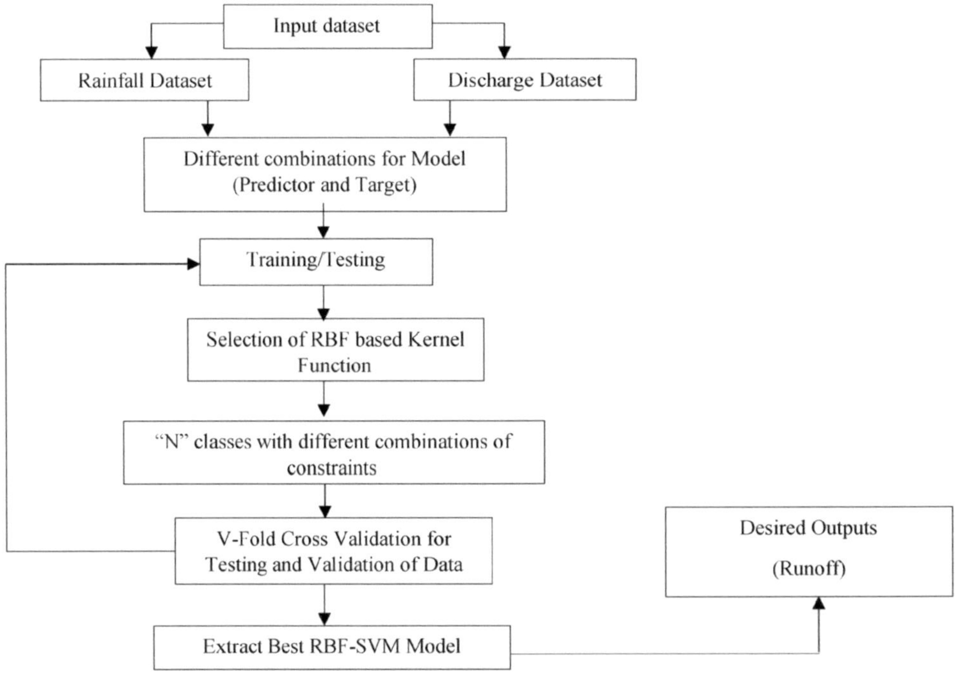

Figure 2.
Working layout of RBF based kernel support vector machine.

2. Applying the RBF kernel function further utilizes the V-fold cross validation to determine the best constraint (penalty constraint).

$$\mathrm{Min}(\sigma)\ \mathrm{J}(\sigma) = 1 - (\sigma) + \mathrm{b}(\sigma) \tag{5}$$

Based on [66] theory of RBF kernel-based SVM, the technique is employed in this research for rainfall-runoff modeling.

2.4 M5 model tree

In the M5 model tree machine learning technique, the following principle converted the space into the area and made the linear regression model. The model's outcome is shown in the modular model, committee machine, with linear models specially designed on appropriate subsets of input space. This design is not innovative. Fusion of specialized technique ("local" model) is passed down in modeling. The finding can clear analogy among Model Trees (MTs), and a combo of linear models utilized in dynamic hydrology since the 1970s- evident paper on multilinear techniques is by [62]. Model tree M5, based on the information theory principle, will have divided multi-dimensional space and create the models automatically based on quality criterion. The number of models can also be varying in number. Computational intelligence techniques combined the numerous models and possibly the combination theory and data-driven outcomes are supporters in hydrology. (example [63], in the fuzzy system, combined hydrological techniques). Computational requirement for model tree raises rapidly with dimensionality [34]. Model tree tackles the task efficiently with high dimension-up to hundreds of attributes. The main advantage of tree models instead of the regression model is that they are smaller than regression trees. The strength of the decision is clear, and regression parameters do not normally involve various variables. M5 algorithm is used for inducing a model tree, which works as shown in.

Suppose collection T of example training is available. Each example is categorized by the values of non-variable set of attributes and has target value. Goal is to build a model with associated target values of training and their input attributes. The efficiency of the model will be calculated by the accuracy, which is forecasting that targets unknown cases shown in **Figure 3**.

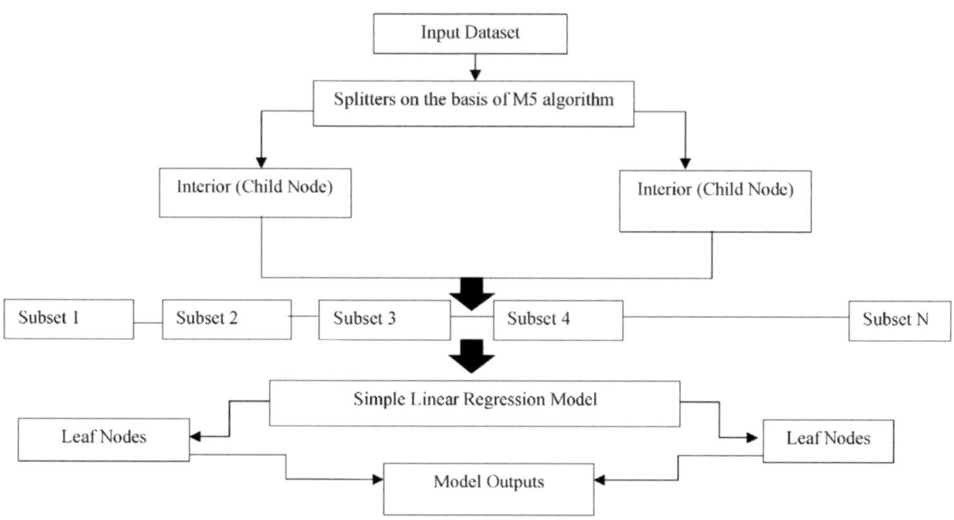

Figure 3.
Working layout of M5 model tree.

2.5 Model performance

Different performance evaluation criteria were used to evaluate the reliability of AITs of the rainfall-runoff process [22, 64] 1) Co-efficient of determination (R^2) [65]; (2) Normalized root mean square error (NRMSE) [66]; (3) Nash-Sutcliffe Coefficient of efficiency (COE) [67] (4) Mean square error (MSE) [68] were used.

$$R^2 = \frac{n(\Sigma xy) - (\Sigma x)(\Sigma y)}{\sqrt{n[\Sigma x2 - (\Sigma x2)][\Sigma y2 - (\Sigma y2)]}} \tag{6}$$

$$NRMSE = \frac{\sqrt{\frac{\sum_{i=1}^{N}(\mathrm{Qobs-Qpre})^2}{\mathrm{N}}}}{\sigma} \tag{7}$$

$$COE = 1 - \frac{\sum_{i=1}^{N}(Qobs - Qpre)^2}{\sum_{i=1}^{N}(Qobs - Qmean)^2} \tag{8}$$

$$MSE = \frac{\sum_{i=1}^{N}(\mathrm{Qobs} - \mathrm{Qpre})^2}{N} \tag{9}$$

Where, Q_{obs} and Q_{pre} are the observed and predicted flows, respectively, while Q_{mean} is the mean of observed flows. R^2 tells us how the fit line of regression approaches the actual data in regression. Value 1 illustrates that the line efficiently fits the real data.

3. Results and discussions

3.1 Rainfall forecasting

Flow Duration Curves (FDCs) were employed to evaluate the applied AITs against the percent of Time. FDCs for all input combinations (C1, C2, C3, C4, C5, C6 and C7) showed a good relationship with applied AITs in both training and testing seasons. To understand the behavior of applied AITs with the Jhelum River basin, the FDCs analysis was executed at nine rainfall stations for the modeling of the rainfall-runoff process as the runoff data was collected from the Mangla reservoir from time duration 1981–2012, the behavior of all techniques necessary to understand throughout the catchment.

The observed hydrographs of low, medium and high percentile flow extracted by the AITs (GEP, RBF-SVM and M5 Model Tree) to access the capability. [52, 69, 70] revealed that the FDCs exposed the relationship between the observed and modeled percentile flow and exceedance probability in the designated time duration. From 1 to 10%, the flow is considered high, 11–89% the flow is medium while, 90–100% the flow is referred to as low flows, which can be clearly seen from.

Furthermore, the percentile flows from 11 to 49%, and 50–89% are considered high medium and low medium flows. The outcomes of FDCs exposed that the GEP was better AIT for high flows and medium-high flows, and it better bonds with FDC of observed flow. Whereas the FDC of the M5 Model Tree better bonds with medium-low and low percentile flows. While RBF-SVM better bonded with the FDCs of low percentile flows. GEP was compared to other AITs was found more accurate DDM and found highly efficient. **RBF-SVM** these trends are shown in **Figure 4**.

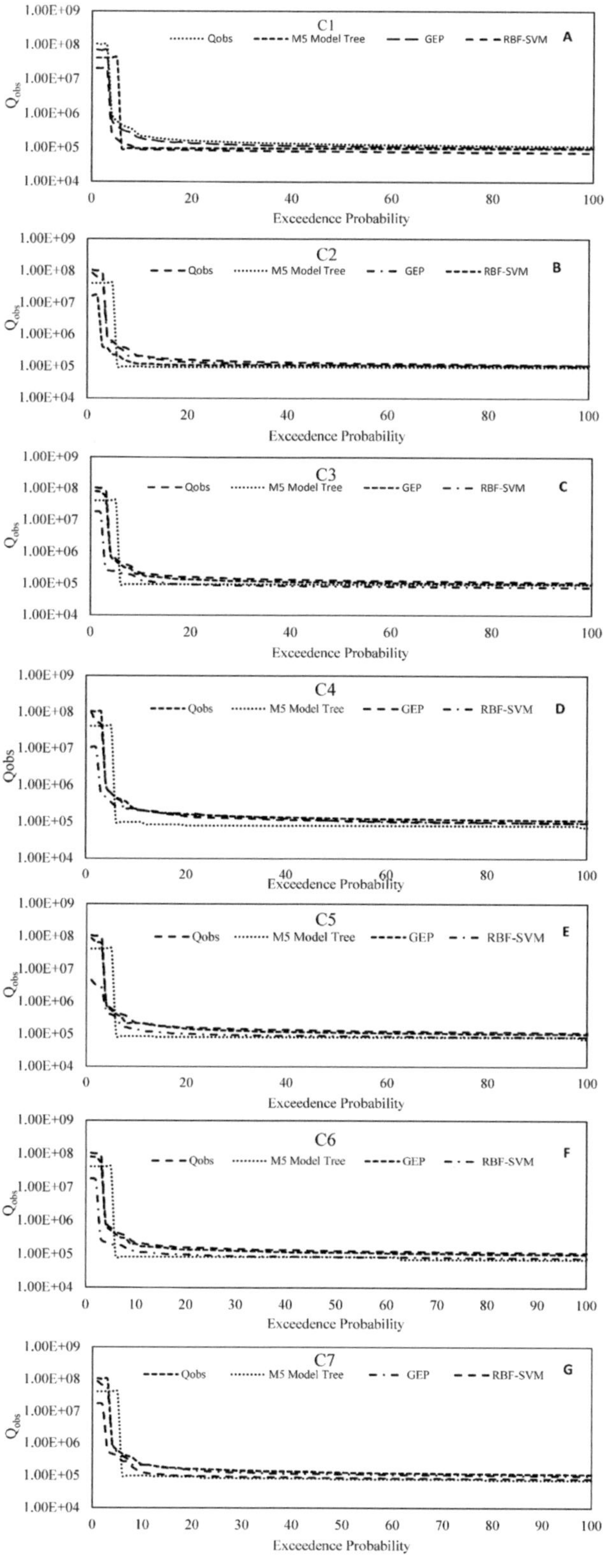

Figure 4.
Flow duration curve (FDC) of observed and simulated daily streamflow in all rivers for various combination C1, C2, C3, C4, C5, C6 and C7 are labeled as A, B, C, D, E, F, and G respectively of Mangla watershed for the time periods 1981–2012.

In RBF kernel-based SVM modeling, the functionality and importance of input combinations were achieved by adjusting the model parameters Gamma, C and P. In other words, the successful application of the RBF-SVM model dependent on accurate determinations of these model parameters. **Figure 5** and **Table 3** show the output results of different input combinations regarding model evaluation performance criterion. It can be clearly seen that RBF-SVM has potential and explicit good performances in training and testing durations of rainfall-runoff modeling. Furthermore, all input combinations employed in this research showed good performance. R^2, COE, MSE and NRMSE for the training period were found 0.99, 1.00, 21245.92 and 820420.17m^3/sec with input C3 and 0.99, 1.00, 21475.00 and 825413.21 m^3/sec respectively with input C6. But input combinations C2 and C4 were found poor combination during training of model with results 0.16, 1.00, −16623.59, 833046.88 m^3/sec and 0.11, 1.00, 980.10, 988371.24 m^3/sec respectively. The behavior of RBF-SVM found poor in both cases due to which showed deprived results. By examining the model evaluation parameters in testing periods, it can be seen that the RBF-SVM model performed and obtain better prediction accuracy. R^2, COE, MSE and NRMSE for the testing duration were found 1.00, 1.00, 188.52 and 1437.96 m^3/sec with C1 and 1.00, 1.00, 147.81 and 1128.49 m^3/sec with input C5, respectively.

.

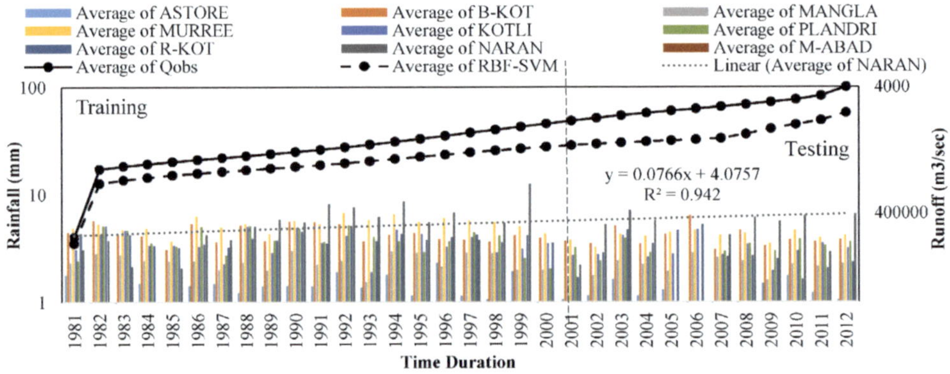

Figure 5.
Hydrographs of RBF-SVM model for overall training and testing rainfall and runoff datasets of Mangla watershed.

	Input Combinations/Model Fitness Criteria	C1	C2	C3	C4	C5	C6	C7
Training	R^2	0.97	0.16	**0.99**	0.16	0.11	0.99	0.98
	COE	1.00	1.00	**1.00**	1.00	1.00	1.00	1.00
	MSE	10366	−8311	**10622**	−1703	490	10737	10409
	NRMSE	401115	416523	**410210**	452420	494185	412706	414639
Testing	R^2	1.00	1.00	1.00	1.00	**1.00**	1.00	1.00
	COE	1.00	1.00	1.00	1.00	**1.00**	1.00	1.00
	MSE	94	131	154	121	**73**	137	96
	NRMSE	718	654	691	588	**564**	613	555

The bold values shows efficient results of model evaluation parameters.

Table 3.
Training and testing outcomes of statistics of RBF-SVM model with different input combinations.

Training	Input	C1	C2	C3	C4	C5	C6	C7
	R^2	0.70	**0.71**	**0.71**	0.65	0.65	0.65	0.65
	COE	1.00	**1.00**	**1.00**	1.00	1.00	1.00	1.00
	MSE	0.00	**0.00**	**0.00**	0.00	0.00	0.00	0.00
	NRMSE	378605	**378579**	**378552**	378596	378596	378570	378570
Testing	R^2	1.00	1.00	0.99	1.00	0.99	0.99	**1.00**
	COE	1.00	1.00	1.00	1.00	1.00	1.00	**1.00**
	MSE	0.00	0.00	0.00	0.00	0.00	0.00	**0.00**
	NRMSE	719	662	683	555	612	698	**443**

The bold values shows efficient results of model evaluation parameters.

Table 4.
Training and testing outcomes of statistics of M5 model tree with different input combinations.

3.2 M5 model tree

The outcomes of training and testing of the M5 Model Tree for the rainfall-runoff process confirms the fact that it has the potential of identifying the relationship between both hydrological variables of a catchment. This statement was confirmed by the model evaluation criteria with low values of NRMSE and high values of R^2 and COE for the validation and testing of the dataset, which suggests the best model fit. The visualization of **Table 4** shows that the M5 Model Tree has the capability to reproduced well by the model with different rainfall-runoff input combinations. The training results indicate that the prediction of Q(t-2) and Q(t-3) quite well for the rainfall-runoff process having results of R^2, COE, MSE and NRMSE, 0.71, 1.00, 0.00, 757158.18 m^3/sec and 0.71, 1.00, 0.00, 757158.18 m^3/sec respectively. During testing of the model, the model evaluation parameters R^2, COE, MSE and NRMSE results are found as 1.00, 1.00, 0.00, 887.52 m^3/sec with input C7, which means that the M5 model tree explicit good results in testing with both rainfall and runoff combinations. The modeling error for the verification of the results indicates high values of R^2 and COE and low values of NRMSE, demonstrating the good M5 model tree performance.

4. Conclusion

The study compared AITs (RBF-SVM and M5 model tree) to understand the rainfall-runoff process in the Jhelum River Basin. Different rainfall and runoff dataset combinations were used to train and test AITs. After training and testing, modeled runoff and observed data was evaluated using R^2, NRMSE, COE and MSE. The conclusion of this study as following:

- Different datasets were analyzed to achieve the target, such as C1, C2, C3, C4, C5, C6 and C7 with lagged past daily rainfall and runoff. Overall, C3 and C7 were performed best among all the dataset. These two datasets showed efficient and accurate results in the training and testing phases.

- The M5 model tree was performed better than other applied techniques. GEP has also exhibited good results to understand the rainfall runoff process.

The RBF-SVM performed less accurate as compared to other applied techniques.

- Flow duration curve (FDCs) were used to compare the modeled and observed dataset of the Jhelum River basin. For High flow and medium-high flows, GEP exhibited well. M5 model tree displayed better results for medium-low and low percentile flows. RBF-SVM exhibited better for low percentile flows. GEP was found the accurate and highly efficient DDM among the AITs applied techniques.

- This study will help understand the complex rainfall-runoff process, which is a stochastic process. Streamflow, weather forecasting plays a key role in water resources management and planning.

Acknowledgements

The dataset of metrological provided by Pakistan metrological department as well as stream flow data provided by WAPDA.

Conflict of interest

The authors hereby declare that there is no conflict of interest.

Author details

Muhammad Waqas[1], Muhammad Saifullah[1]*, Sarfraz Hashim[1], Mohsin Khan[1] and Sher Muhammad[2]

1 Department of Agricultural Engineering, Muhammad Nawaz Sahreef, University of Agriculture, Multan, Pakistan

2 International Center for Integrated Mountain Development (ICIMOD), Kathmandu, Nepal

*Address all correspondence to: muhammad.saifullah@mnsuam.edu.pk

References

[1] Morley, S.J.S.W., *Challenges and opportunities in magnetospheric space weather prediction.* 2020. **18**(3): p. e2018SW002108.

[2] Wagena, M.B., et al., *Comparison of short-term streamflow forecasting using stochastic time series, neural networks, process-based, and Bayesian models.* 2020. **126**: p. 104669.

[3] López-Felices, B., et al., *Contribution of irrigation ponds to the sustainability of agriculture.* A Review of Worldwide Research. 2020. **12**(13): p. 5425.

[4] O'Keeffe, J., et al., Isolating the impacts of anthropogenic water use within the hydrological regime of north India. 2020. **45**(5): p. 1217-1228.

[5] Niu, W.-j., Z.-k.J.S.C. Feng, and Society, *Evaluating the performances of several artificial intelligence methods in forecasting daily streamflow time series for sustainable water resources management.* 2021. **64**: p. 102562.

[6] Shoaib, M., et al., Input Selection of Wavelet-Coupled Neural Network Models for Rainfall-Runoff Modelling. 2019. **33**(3): p. 955-973.

[7] Penman, H.J.W., *Weather, plant and soil factors in hydrology.* 1961. **16**(7): p. 207-219.

[8] Peel, M.C. and T.A.J.W.I.R.W. McMahon, *Historical development of rainfall-runoff modeling.* 2020. 7(5): p. e1471.

[9] Dariane, A.B. And F.J.W.r.m. Karami, *Deriving hedging rules of multi-reservoir system by online evolving neural networks.* 2014. **28**(11): p. 3651-3665.

[10] Ruelland, D.J.H. and E.S. Sciences, *Should altitudinal gradients of temperature and precipitation inputs be inferred from key parameters in snow-hydrological models?* 2020. **24**(5): p. 2609-2632.

[11] Smith, J., R.N.J.J.o.w.r.p. Eli, and management, *Neural-network models of rainfall-runoff process.* 1995. **121**(6): p. 499-508.

[12] Hsu, K.l., H.V. Gupta, and S.J.W.r.r. Sorooshian, Artificial neural network modeling of the rainfall-runoff process. 1995. **31**(10): p. 2517-2530.

[13] Dawson, C.W. and R.J.H.S.J. Wilby, *An artificial neural network approach to rainfall-runoff modelling.* 1998. **43**(1): p. 47-66.

[14] Tokar, A.S. And P.a.J.J.o.H.E. Johnson, *Rainfall-runoff modeling using artificial neural networks.* 1999. **4**(3): p. 232-239.

[15] Tayfur, G., T. Moramarco, And V.P.J. H.P.a.I.J. Singh, *Predicting and forecasting flow discharge at sites receiving significant lateral inflow.* 2007. **21**(14): p. 1848-1859.

[16] Kişi, Ö.J.J.o.H.E., *Neural networks and wavelet conjunction model for intermittent streamflow forecasting.* 2009. **14**(8): p. 773-782.

[17] Engineering, A.T.C.o.A.o.A.N.N.i.H. J.J.o.H., *Artificial neural networks in hydrology. I: Preliminary concepts.* 2000. **5** (2): p. 115-123.

[18] Engineering, A.T.C.o.A.o.A.N.N.i.H. J.J.o.H., *Artificial neural networks in hydrology. II: Hydrologic applications.* 2000. **5**(2): p. 124-137.

[19] Campolo, M., P. Andreussi, And a.J. W.r.r. Soldati, *River flood forecasting with a neural network model.* 1999. **35**(4): p. 1191-1197.

[20] Coulibaly, P., F. Anctil, And B.J.J.o. H. Bobée, *Daily reservoir inflow*

forecasting using artificial neural networks with stopped training approach. 2000. **230**(3–4): p. 244-257.

[21] Cigizoglu, H.K. and Ö.J.H.R. Kişi, *Flow prediction by three back propagation techniques using k-fold partitioning of neural network training data.* 2005. **36** (1): p. 49-64.

[22] Shoaib, M., et al., *Runoff forecasting using hybrid wavelet gene expression programming (WGEP) approach.* 2015. **527**: p. 326-344.

[23] Yaseen, Z.M., et al., *Artificial intelligence based models for stream-flow forecasting: 2000–2015.* Journal of Hydrology, 2015. **530**: p. 829-844.

[24] Vapnik, V.J.I., New York, Statistical Learning Theory. John Wiley&Sons. 1998.

[25] Bray, M. And D.J.J.o.H. Han, *Identification of support vector machines for runoff modelling.* 2004. **6**(4): p. 265-280.

[26] Suykens, J.A. *Nonlinear modelling and support vector machines*. in *IMTC 2001. proceedings of the 18th IEEE instrumentation and measurement technology conference. Rediscovering measurement in the age of informatics (Cat. No. 01CH 37188)*. 2001. IEEE.

[27] Van Gestel, T., et al., Benchmarking least squares support vector machine classifiers. 2004. **54**(1): p. 5-32.

[28] Espinoza, M., et al., Electric load forecasting. 2007. **27**(5): p. 43-57.

[29] Dutta, S., et al., *Machine learning algorithms and their application to ore reserve estimation of sparse and imprecise data.* 2010. **2**(02): p. 86.

[30] Samsudin, R., A. Shabri, And P.J.J.o. a.s. Saad, *A comparison of time series forecasting using support vector machine and artificial neural network model.* 2010. **10**(11): p. 950-958.

[31] Dibike, Y.B., et al., Model induction with support vector machines: introduction and applications. 2001. **15** (3): p. 208-216.

[32] Tripathi, S., V. Srinivas, And R.S.J.J. o.h. Nanjundiah, *Downscaling of precipitation for climate change scenarios: a support vector machine approach.* 2006. **330**(3–4): p. 621-640.

[33] Chen, Y.-W. and C.-J. Lin, *Combining SVMs with various feature selection strategies*, in *Feature extraction*. 2006, Springer. p. 315-324.

[34] Quinlan, J.R. *Learning with continuous classes*. in *5th Australian joint conference on artificial intelligence*. 1992. World Scientific.

[35] Rashidi, S., et al., *Evaluating the support vector machine for suspended sediment load forecasting based on gamma test.* 2016. **9**(11): p. 583.

[36] Sarangi, A. And a.J.a.w.m. Bhattacharya, *Comparison of artificial neural network and regression models for sediment loss prediction from Banha watershed in India.* 2005. **78**(3): p. 195-208.

[37] Solomatine, D.P. and K.N.J.H.S.J. Dulal, *Model trees as an alternative to neural networks in rainfall—runoff modelling.* 2003. **48**(3): p. 399-411.

[38] Solomatine, D.P. and Y.J.J.o.H.E. Xue, *M5 model trees and neural networks: application to flood forecasting in the upper reach of the Huai River in China.* 2004. **9**(6): p. 491-501.

[39] Ghaemi, A., et al., *On the applicability of maximum overlap discrete wavelet transform integrated with MARS and M5 model tree for monthly pan evaporation prediction.* 2019. **278**: p. 107647.

[40] Singh, T., et al., *Modeling oblique load carrying capacity of batter pile groups using neural network, random forest regression and M5 model tree.* 2019. **13**(3): p. 674-685.

[41] García-Nieto, P., et al., *Modeling algal atypical proliferation using the hybrid DE–MARS–based approach and M5 model tree in La Barca reservoir: A case study in northern Spain.* 2019. **130**: p. 198-212.

[42] Ali, A. and M.J.L. Imran, *National Spatial Data Infrastructure vs.* Cadastre System for Economic Development: Evidence from Pakistan. 2021. **10**(2): p. 188.

[43] Qazi, N.Q., et al., *Hydrology of the Himalayas*, in *Himalayan Weather and Climate and their Impact on the Environment*. 2020, Springer. p. 419-450.

[44] Grover, S., et al., Modeling Hydrological Processes in Ungauged Snow-Fed Catchment of Western Himalaya. 2020. **47**(6): p. 987-995.

[45] Zaman, M., et al., *Event-Based Time Distribution Patterns, Return Levels, and Their Trends of Extreme Precipitation across Indus Basin.* 2020. **12**(12): p. 3373.

[46] Haider, H., et al., Appraisal of Climate Change and Its Impact on Water Resources of Pakistan: A Case Study of Mangla Watershed. 2020. **11** (10): p. 1071.

[47] Bastiaanssen, W., et al., *Surface energy balance and actual evapotranspiration of the transboundary Indus Basin estimated from satellite measurements and the ETLook model.* 2012. **48**(11).

[48] Fowler, H. And D.J.J.o.C. Archer, *Conflicting signals of climatic change in the Upper Indus Basin.* 2006. **19**(17): p. 4276-4293.

[49] Young, G. and K.J.I.P. Hewitt, *Hydrology research in the upper Indus basin,* Karakoram Himalaya, Pakistan. 1990. **190**: p. 139-152.

[50] Immerzeel, W.W., et al., Large-scale monitoring of snow cover and runoff simulation in Himalayan river basins using remote sensing. 2009. **113** (1): p. 40-49.

[51] Bookhagen, B. and D.W.J.G.R.L. Burbank, *Topography, relief, and TRMM-derived rainfall variations along the Himalaya.* 2006. **33**(8).

[52] Archer, D.R. And H.J.J.o.H. Fowler, *Using meteorological data to forecast seasonal runoff on the River Jhelum, Pakistan.* 2008. **361**(1–2): p. 10-23.

[53] Azmat, M.J.P.P.D.T., University of Porto, Porto, Portugal, *Water resources availability and hydropower production under current and future climate scenarios: The case of Jhelum River Basin, Pakistan.* 2015.

[54] Azmat, M., et al., *Impacts of changing climate and snow cover on the flow regime of Jhelum River,* Western Himalayas. 2017. **17**(3): p. 813-825.

[55] Ramsey, F.L.J.T.A.o.S., *Characterization of the partial autocorrelation function.* 1974. **2**(6): p. 1296-1301.

[56] van Heel, M., M. Schatz, And E.J.U. Orlova, *Correlation functions revisited.* 1992. **46**(1–4): p. 307-316.

[57] Shiri, J., et al., *Daily reference evapotranspiration modeling by using genetic programming approach in the Basque Country (Northern Spain).* 2012. **414**: p. 302-316.

[58] Noori, R., et al., *Assessment of input variables determination on the SVM model performance using PCA,* Gamma test, and forward selection techniques for

monthly stream flow prediction. 2011. **401**(3–4): p. 177-189.

[59] Li, C.-H., et al. *An automatic method for selecting the parameter of the normalized kernel function to support vector machines*. in *2010 International Conference on Technologies and Applications of Artificial Intelligence*. 2010. IEEE.

[60] Li, C.-H., et al. *An automatic method for selecting the parameter of the RBF kernel function to support vector machines*. in *2010 IEEE International Geoscience and Remote Sensing Symposium*. 2010. IEEE.

[61] Kuo, B.-C., et al., A kernel-based feature selection method for SVM with RBF kernel for hyperspectral image classification. 2013. 7(1): p. 317-326.

[62] Becker, A. and Z.W.J.W.R.R. Kundzewicz, *Nonlinear flood routing with multilinear models*. 1987. **23**(6): p. 1043-1048.

[63] Xiong, L., A.Y. Shamseldin, And K. M.J.J.o.h. O'connor, *A non-linear combination of the forecasts of rainfall-runoff models by the first-order Takagi–Sugeno fuzzy system*. 2001. **245**(1–4): p. 196-217.

[64] Shoaib, M., et al., A wavelet based approach for combining the outputs of different rainfall–runoff models. 2018. **32**(1): p. 155-168.

[65] Menard, S.J.T.A.S., *Coefficients of determination for multiple logistic regression analysis*. 2000. **54**(1): p. 17-24.

[66] Levinson, N.J.J.o.M. and Physics, *The Wiener (root mean square) error criterion in filter design and prediction*. 1946. **25**(1-4): p. 261-278.

[67] Nash, J.E. And J.V.J.J.o.h. Sutcliffe, *River flow forecasting through conceptual models part I—A discussion of principles*. 1970. **10**(3): p. 282-290.

[68] Allen, D.M.J.T., *Mean square error of prediction as a criterion for selecting variables*. 1971. **13**(3): p. 469-475.

[69] Babur, M., et al., *Assessment of climate change impact on reservoir inflows using multi climate-models under RCPs—The case of Mangla Dam in Pakistan*. 2016. **8**(9): p. 389.

[70] Hayat, H., et al., Simulating Current and Future River-Flows in the Karakoram and Himalayan Regions of Pakistan Using Snowmelt-Runoff Model and RCP Scenarios. 2019. **11**(4): p. 761.

Chapter 3

Application of Kalman Filter and Breeding Ensemble Technique to Forecast the Tropical Cyclone Activity

Cong Thanh, Dao Nguyen Quynh Hoa and Tran Tan Tien

Abstract

Tropical cyclone (TC) is one of the major meteorology disasters, as they lead to deaths, destroy the infrastructure and the environment. Therefore, how to improve the predictability of TC's activities, such as formation, track, and intensity, is very important and is considered an important task for current operational predicting TC centers in many countries. However, predicting TC's activities has remained a big challenge for meteorologists due to our incomplete understanding of the multiscale interaction of TCs with the ambient environment and the limitation of numerical weather forecast tools. Hence, this chapter will exhibit some techniques to improve the ability to predict the formation and track of TCs using an ensemble prediction system. Particularly, the Local Ensemble Transform Kalman Filter (LETKF) scheme and its implementation in the WRF Model, as well as the Vortex tracking method that has been applied for the forecast of TCs formation, will be presented in subSection 1. Application of Breeding Ensemble to Tropical Cyclone Track Forecasts using the Regional Atmospheric Modeling System (RAMS) model will be introduced in subSection 2.

Keywords: The WRF-LETKF system, Ensemble forecast technique, Breeding Ensemble, data assimilation system, Tropical cyclone forecast

1. Introduction

1.1 The forecast of TCs formation using the ensemble Kalman filter

Among several approaches for real-time monitoring and forecasting of TC formation, direct numerical products from global and regional weather prediction models appear to be the most reliable at present, despite their inherent limitations and uncertainties (e.g., see [1, 2]). The skillful performance of TC formation forecasts by numerical models has been well documented in many previous studies [3–8]. This achievement of numerical models is attributed to a variety of advanced research on upgrading parameterizations of physics, resolution, computational resources, and data assimilation schemes [1, 9]. Among several different assimilation schemes, the ensemble Kalman filter (EnKF) has been extensively applied to many practical problems in recent years due to its straightforward implementation for TC forecast applications [10–16]. The use of EnKF for TC forecasting applications is increasingly popular, given the current availability of real-time flight reconnaissance data that allows direct assimilation of airborne observations without the need of a bogus vortex (e.g., see [10, 14–16]). In essence, the development of EnKF addressed the problem when using variational assimilation schemes in which the background covariance matrix is allowed to be time-dependent. Hence, the model would adapt better with fast-evolving and complicated dynamical systems such as TCs or mesoscale convective systems [12, 17–19]. There is an efficient method of implementing an Ensemble Kalman Filter (EnKF), which was called a Local Ensemble Transform Kalman Filter (LETKF) scheme.

1.2 Data assimilation system

In this section, the LETKF algorithm proposed by Ott et al. [20] and Hunt et al. [21] is adopted and implemented for the WRF Model. The primary usage of the LETKF algorithm is utilizing the background ensemble matrix as an operator to transform state vectors from a model space spanned by the model grid points within a local patch to an ensemble space spanned by ensemble members. The procedures for calculating matrix and generating the ensemble analyses are executed in this low dimension ensemble space at every single grid point. In this sense, the LETKF scheme allows the ensemble space to be performed locally and in parallel efficiently for practical problems, especially when carrying out a large-volume of data (e.g., see [8, 11, 12, 22–25]).

With its promising capability, LETKF has been implemented in the WRF Model (V3.6, hereafter referred to as the WRF-LETKF system). With an aim to practical forecasting applications, all the observations utilizing in the WRF – LETKF scheme are preprocessing in a quality control taken by the WRF data assimilation (WRFDA) component. In addition, the WRFDA component also generates lateral boundary conditions for each ensemble member once obtained the analysis update. Hence, each ensemble member possesses its own boundary dynamically consistent with its own updated initial conditions. More details in the WRF-LETKF design can be found in [12, 24]. The focal point here is how the ensembles with and without augmented observations perform. In this regard, the relative differences in the output among these ensembles can derive the main effects of additional augmented observations.

To begin the ensemble system, a first-guess background is generated in a cold-start ensemble by first using 3DVAR scheme to produce an analysis from a GFS initial condition. Random perturbations with standard deviations of 1 ms^{-1} for the wind field, 1 K for temperature, and 1×10^{-3} $kgkg^{-1}$ for specific humidity at all model grid points are then added to the 3DVAR-generated analyses for the cold-start ensemble. The 3DVAR-generated analyses as initial conditions for 12-h running in a manner that the outputs from these 12-h integrations can be subsequently used as a *warm-start* background for the LETKF ensemble assimilation in the next cycle. Note that these random perturbations are added only for the first cold-start cycle to create a background ensemble. All subsequent warm-run cycles use the WRF-LETKF 12-h forecasts as a background ensemble and so no additional random noises are necessary. The newly generated analysis perturbation ensemble at each cycle is then added to the GFS analysis to produce the next ensemble initial conditions when run in the cycling mode as described in [26].

1.3 The LETKF algorithm

To get a better understanding of the LETKF algorithm mentioned in the previous sub-section. A brief description of this LETKF algorithm that developed by Kieu et al. [12] has been presented below:

Assume that give a background ensemble $\{\mathbf{x}^{b(i)}: i = 1, 2, \ldots, k\}$, where k is the number of ensemble members (assuming that the analysis is taken one at a time, so the time index is not included). According to Hunt et al. [21], an ensemble mean $\bar{\mathbf{x}}^b$ and an ensemble perturbation matrix $\mathbf{X}^b$ are defined respectively as:

$$\bar{\mathbf{x}}^b = \frac{1}{k}\sum_{i=1}^{k}\mathbf{x}^{b(i)}.$$

$$\mathbf{X}^b = \left\{\mathbf{x}^{b(1)} - \bar{\mathbf{x}}^b, \mathbf{x}^{b(2)} - \bar{\mathbf{x}}^b, \ldots, \mathbf{x}^{b(k)} - \bar{\mathbf{x}}^b\right\}. \tag{1}$$

Let $\mathbf{x} = \bar{\mathbf{x}}^b + \mathbf{X}^b\mathbf{w}$, where w is a local vector in the ensemble space, the local cost function to be minimized in the ensemble space is given by:

$$\hat{j}(\mathbf{w}) = (k-1)\mathbf{w}^T\left\{\boldsymbol{I} - \left(\mathbf{X}^b\right)^T\left[\mathbf{X}^b\left(\mathbf{X}^b\right)^T\right]^{-1}\mathbf{X}^b\right\}\mathbf{w} + J\left[\mathbf{x}^b + \mathbf{X}^b\mathbf{w}\right], \tag{2}$$

Where $J[\mathbf{x}^b + \mathbf{X}^b\mathbf{w}]$ is the cost function in the model space. If one defines the null space of $\mathbf{X}^b$ as $N = \{\mathbf{v}|\mathbf{X}^b\mathbf{v} = 0\}$, then the cost function $\hat{\boldsymbol{J}}(\mathbf{w})$ is divided into two parts: one containing the component of **w** in *N* (the first term in Eq. (2)), and the second depending on the components of **w** that are orthogonal to *N*. By requiring that the mean analysis state $\bar{\mathbf{w}}^a$ is orthogonal to *N* such that the cost function $\hat{\boldsymbol{J}}(\mathbf{w})$ is minimized, the mean analysis state and its corresponding analysis error covariance matrix in the ensemble space can be found as:

$$\bar{\mathbf{w}}^a = \hat{\mathbf{P}}^a\left(\mathbf{Y}^b\right)^T\mathbf{R}^{-1}\left[\mathbf{y}^0 - \boldsymbol{H}\left(\bar{\mathbf{x}}^b\right)\right] \tag{3}$$

$$\hat{\mathbf{P}}^a = \left[(k-1)\mathbf{I} + \left(\mathbf{Y}^b\right)^T\mathbf{R}^{-1}\mathbf{Y}^b\right]^{-1}, \tag{4}$$

Where $\mathbf{Y}^b \equiv \boldsymbol{H}(\mathbf{x}^{b(i)} - \bar{\mathbf{x}}^b)$ is the ensemble matrix of background perturbations valid at the observation locations, and **R** is the observational error covariance matrix. By noting that the analysis error covariance matrix $\mathbf{P}^a$ in the model space and $\hat{\mathbf{P}}^a$ in the ensemble space have a simple connection of $\mathbf{P}^a = \mathbf{X}^b\hat{\mathbf{P}}^a\left(\mathbf{X}^b\right)^T$, the analysis ensemble perturbation matrix $\mathbf{X}^a$ can be chosen as follows:

$$\mathbf{X}^a = \mathbf{X}^b\left[(k-1)\hat{\mathbf{P}}^a\right]^{1/2}. \tag{5}$$

The analysis ensemble $\mathbf{x}^a$ is finally obtained as:

$$\mathbf{x}^{a(i)} = \bar{\mathbf{x}}^b + \mathrm{X}^b\left\{\bar{\mathbf{w}}^a + \left[(k-1)\hat{\mathbf{P}}^{a(i)}\right]^{1/2}\right\}. \tag{6}$$

Detailed handling of more general nonlinear and synchronous observations in LETKF can be found in [21]. It should be noticed that the above formulas are only valid without model errors. To take into account the model errors, Hunt et al. [21] suggested that a multiplicative factor should be introduced in Eq. (4) (specifically, the first factor on the right hand side of Eq. (4)). This simple additional multiplicative inflation is easy to implement in the scheme, and has been shown to be efficient in many applications of the LETKF (e.g., see [25, 27, 28]).

1.4 Vortex tracking method

Constructing a suitable vortex-tracking algorithm is a must-have procedure to detect the formation of a newly developed TC center in all ensemble members since their outputs are diverse. A good detection scheme allows one to define and verify the location and the timing of TC formation centers. This step is crucial in every TC formation study, due to the difficulties in capturing the incoherent structure of tropical cyclones at the early genesis stages. More precisely, one cannot apply general criteria such as a midlevel warm-core anomaly, maximum vorticity center for the tropical disturbances as for mature TCs. Instead, the early formation of a tropical depression is often imprinted by the existence of an upper-level cold core and/or a weak surface low pressure rather than a midlevel warm core (see, e.g., [29, 30]). Thus, very few conditions can be practically applied to detect a formation center during the genesis stage. To detect TC formation centers for real-time forecast, a simple scheme has been built upon standard conditions related to the maximum surface wind and the minimum central pressure, as follows:

First, the minimum sea level pressure P_{min} within the study area is searched at every model grid point of each ensemble member output at each forecast lead time. Any location with $P_{min} < 1004$ hPa will be noted down as a potential candidate for TC formation location at that forecast lead time for that particular ensemble member.

Second, once a possible location of TC formation is defined, the maximum 10-m wind speed Vmax in an area of $4^0 \times 4^0$ surrounding the minimum pressure center is checked and recorded. A TC formation center will be marked if the condition $V_{max} \geq 10$ ms^{-1} is satisfied. It is noteworthy that this value is considerably smaller than the global definition of a tropical depression wind speed ($\sim$17 ms^{-1}), due to the relatively coarse 27−/9-km resolution configuration of WRF-LETKF system. Visualizing verification of each TC circulation center detected based on this threshold proves that these criteria can properly identify the center of tropical cyclone like vortex during the genesis stage. Therefore, this threshold for V_{max} is used for all genesis analyses. In fact, these criteria of tracking TC formation centers are somewhat intuitive and require further verification. However, this approach is acceptable in evaluating the augmented observational data impacts on TC formation forecasts among ensemble forecasts. As long as the tracking scheme remains certain in all analyses, the comparison of TC formation forecasts should answer the question about the performance of augmented observations in ensemble forecasts.

1.4.1 Example 1

The WRF-LETKF (WRF V3.6) system has been applied to study the formation of Typhoon Wutip. With target is to evaluate the sensitivity of TC formation forecast to different types of augmented observations. The WRF-LETKF system is designed in such a way that all observations are subject to quality control by the WRF data assimilation (WRFDA) component before used by the LETKF algorithm (More details about the implementation of the WRF-LETKF design can be found in [12, 22]. There is a total of 21 ensemble members was made (due to limited computational and storage resources) and all ensemble experiments are integrated for three days starting from 1200 UTC 23 September, which is approximately 48 h before a tropical depression precursor of Wutip was first reported in the TC vital record at 1200 UTC 25 September. The multiple physical schemes have been used in categorizing among ensemble experiments are 1) two cumulus parameterization schemes including the Betts–Miller–Janjic′ (BMJ) cumulus parameterization and the Kain–Fritsch with shallow convection schemes, 2) three planetary boundary layer (PBL) parameterization schemes including the Yonsei University, the Mellor–Yamada–Janjic′, and the simple Medium-Range Forecast (MRF) schemes, 3) three microphysical schemes including the WSM 3 microphysics, the Kessler, and the Lin et al. schemes; and 4) two longwave radiative schemes including the Dudhia and the Goddard schemes for both longwave and shortwave radiations. The cold start cycle is therefore initialized at 0000 UTC 23 September to generate a background ensemble for the first-guess cycle at 1200 UTC 23 September. Afterwards, the subsequent cycles are implemented at every 6 h from 1200 UTC 23 September to 1200 UTC 26 September.

The augmented observational data used in the WRF-LETKF assimilation scheme include two main sources. The first is the satellite data (CIMSS-AMV) derived atmospheric motion vector (AMV) data maintained by the Cooperative Institute for Meteorological Satellite Studies (CIMSS), University of Wisconsin [21, 31–34] due to this data covers a large area where TC genesis may take place. The second source of local augmented observations in the domain of influence to Vietnam's coastal region (DOIV) is also used, including 96 aviation routine weather (METAR) reports from routine scheduled observations, 31 ship/buoy (SHIP/BUOY) station reports, 59 enhanced sounding stations (SOUND), and 404 surface synoptic observations (SYNOP) reports of weather observations during the 0000 UTC 24 September–0000 UTC 27 September period.

Results show critical impacts of the (CIMSS-AMV) data in improving the large – scale environment favorable or hostile to the formation of Typhoon Wutip among ensemble members, which is dynamically controlled by monsoon trough. The results show the optimality of data impacts at cycle 36 h prior to Wutip's observed formation and decrease as forecast cycles are closer to formation period. In contrast, the data assimilation with only surface and local station data proves that these source data are not enough to help describe the strength of monsoon trough due to their scattered distributions (**Figures 1–3**).

By choosing Typhoon Wutip as a case study, it was demonstrated that the initial conditions for tropical cyclogeneses in large-scale monsoon trough environment are sensitive to augmented observations. It could allow a range of outcomes for timing and location predictability of TC formation, especially at 36-hr cycle ensemble. Our results could present the importance of augmented observations, especially the

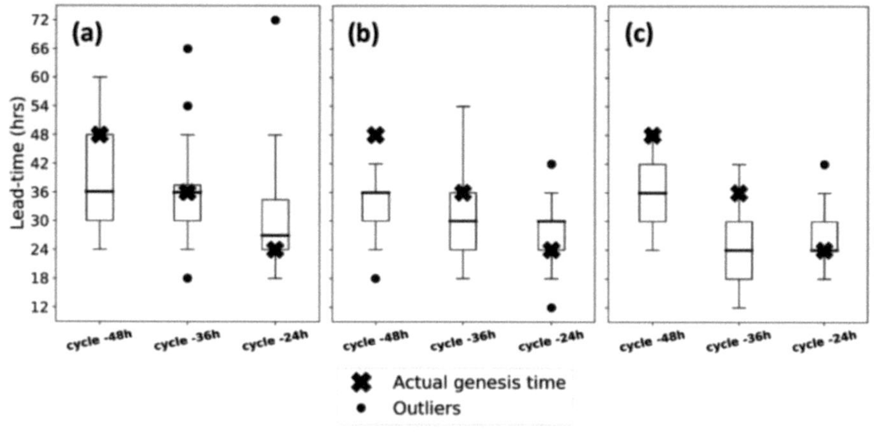

Figure 1.
Boxplots of the timing for Wutip formation for three consecutive cycles 1200 UTC 23 Sep, 0000 UTC 24 Sep, and 1200 UTC 24 Sep, corresponding to 48, 36, and 24 h prior to the formation of Wutip depression for (a) the WRF-LETKF, (b) assimilation without the CIMSS-AMV data (NAMV), and (c) the GFS initial data [hereafter to as no data assimilation (NDA) ensemble]. The bold cross denotes the actual time that Wutip first became a tropical depression at 1200 UTC 25 Sep [35].

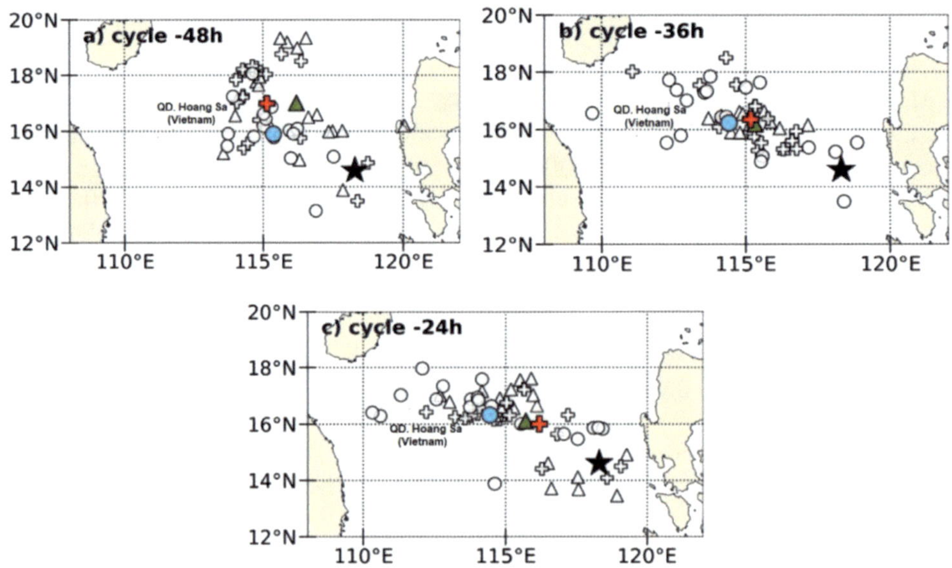

Figure 2.
Distribution of the location of the Wutip's formation centers as forecast by the WRF-LETKF (triangle), the assimilation without the CIMSS-AMV data ensemble (circle), and no data assimilation ensemble (cross) for (a) 48-, (b) 36-, and (c) 24-h cycles. Color symbols denote the ensemble means of corresponding forecasts [35].

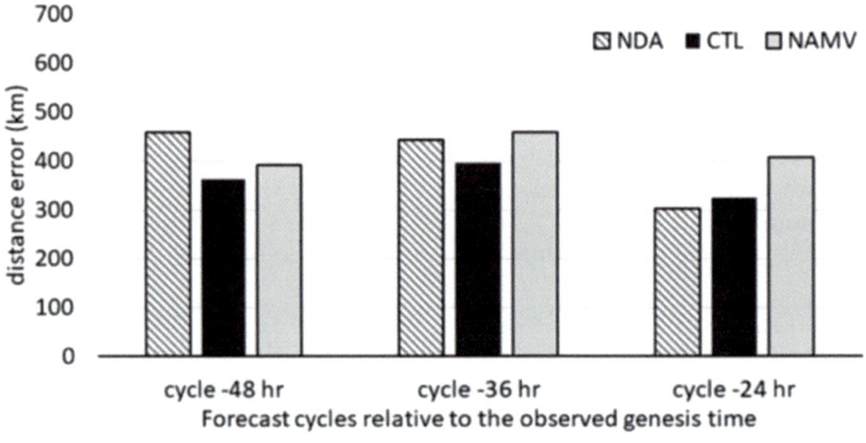

Figure 3.
Ensemble mean distance errors between the forecasted and observed location of Wutip's formation reported at 1200 UTC 25 Sep for three cycles of 48, 36, and 24 h obtained from the WRF-LETKF forecast (black), NDA forecast (stripe), and NAMV forecasts (light shaded) [35].

satellite AMV data, for the prediction of TC formation at certain lead times that are vital for operational TC forecasts. This case study is typical for TC formation in the WPAC basin, but not representative and may not be applied to other tropical cyclogenesis pathways. While WRF-LETKF has been utilized in forecasting tropical cyclogenesis in the marsupial paradigm of African Easterly Wave [36, 37], it has not been focused in the physical mechanisms of TCs formation in the BIEN DONG basin before. Wutip's formation is strongly rooted in the monsoon trough, as most of the tropical cyclones in the BIEN DONG form within this pattern per year. The performance of WRF-LETKF with augmented observations in this case study has innovated to the upcoming studies in more properly general examination when designing future observing systems.

2. Application of breeding ensemble to tropical cyclone track forecasts

2.1 Background

With limited range of predictability at the convective scale and short timescale and the complex variation of nature, predicting tropical cyclone (TC) tracks and intensity is one specific example that demonstrates vividly the sensitivity of numerical models to uncertainties in the atmosphere [26, 38, 39]. The inherent uncertainties associated with our current incomplete understanding of model physical processes or numerical approximations often lead to large errors in track and intensity forecast, especially at the lead times longer 3 days or under circumstances interacting with uneven terrain or complicated vortex mergers [40–42]. Currently, the US Joint Typhoon Warning Center (JTWC) showed that the official track errors in the North Western Pacific (WPAC) basin are as high as 220 km at 3-day estimation and 450 km at 5-day estimation. Likewise, the intensity forecast errors make no headway since no significant update was taken at all forecast ranges during the last 30 years. The recent effort to calculate uncertainty in TC forecasts is based on the ensemble prediction systems. Generally, there are 3 major special techniques to develop an ensemble forecast system include: (1) use the different initial conditions obtained from a posterior analysis error distribution (the Monte-Carlo ensembles) for one specific model, (2) Use a single initial condition for multiple different prediction models; and (3) use combine both dissimilar initial conditions and different prediction models.

The breeding ensemble approach in the first direction was first implemented in the operational Global Forecasting System (GFS) at the National Center for Environmental Prediction (NCEP, by Toth and Kalnay in 1993 and 1997 [43, 44], hereinafter TK93 and TK97, respectively) in 1993, and then became more popular and more applied in practice. The breeding method continuously employed previous cycles to calculate the fastest growing instabilities and then normalized these errors vectors into the so-called the bred vectors. This procedure could allow projecting the fastest growing modes onto the calculated bred vectors in a shade of perturbations in each breeding cycle. Likewise, a similar ensemble forecasting technique generating singular vectors instead of bred vectors is implemented in the European Center for medium-term weather forecast (ECMWF) in early 1992 [45, 46]. Although theoretically, the fastest growing modes should be projected onto bred vectors (at the far limit of the backward Lyapunov vectors), the experimental results retrieved from the TK93's breeding method indicate that the produced TC ensemble tracks could be very similar to each other, i.e., the spread of the system was relatively narrow (**Figure 4**). One possible explanation for such small ensemble dispersion is because the bred vectors collapsed into a similar dominant direction after several cycles, which is not an uncommon issue (e.g., see [48, 49]). The singular vectors display the fastest growing modes in terms of orthogonal directions within a short-range interval (via a tangential linear model). In contrast, the bred vectors are some extent equivalent to the leading Lyapunov vectors in a nonlinear finite-amplitude method [43, 50]. This method allows the bred vectors collapsing afterwards, and becoming linearly independent (non-orthogonal) in the presence of the lower dimension attractor [48, 51].

By consider both the spatial–temporal variations of the scaling vector at each cycle, the bred vectors could capture the local growing directions and thus allow for

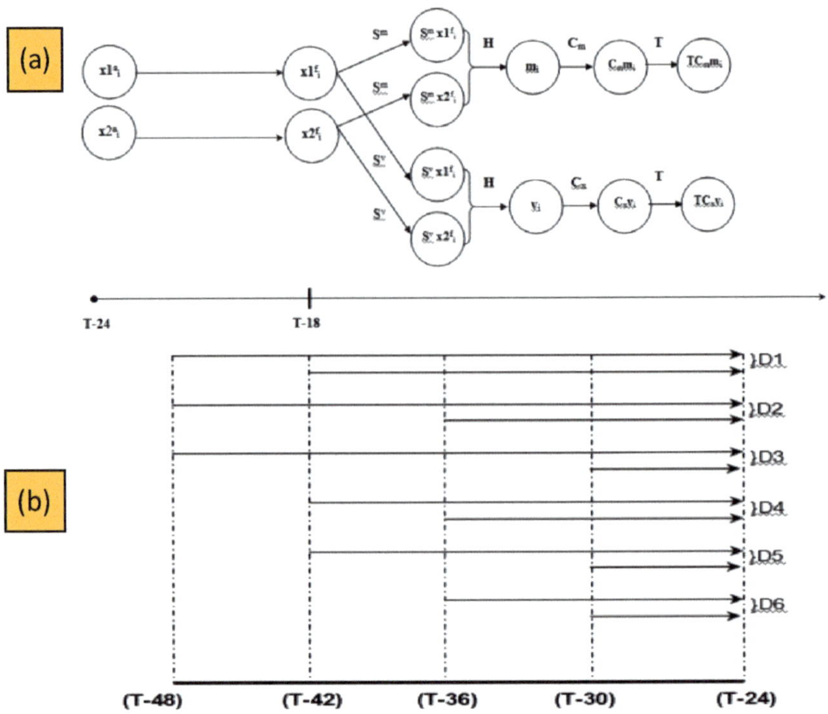

Figure 4.
Schematic design of the TC-breeding ensemble technique: a) illustration of generating environmental bred vectors and TC bred vectors during a warm start cycle (from 24 h to 18 h before the target forecast date) b) illustration of making six pairs of lagged-averaged forecast (LAF) vectors for the first cold start cycle used in the TC-breeding ensemble [47].

larger ensemble spread [44, 52]. However, fast convective instabilities still quickly saturate after several breeding cycles, especially within the region where the atmospheric dynamics are complicated [53, 54]. Unfortunately, the TCs system act as such complicated phenomenon with multi-scale interactions. It is expected that the instability within the storm's inner-core should behave differently as compared to the outer environmental region. Previous studies (see [55– 57] indicated that perturbations inside the TC inner-core area often develop and propagate rapidly in the manner of vortex Rossby waves with typical time scale of 12–24 h. Contrarily, the large-scale environmental related-perturbations propagate in a much smaller time scale, often manifested in terms of gravity waves and mesoscale clustering along the most unstable regions [58]. Representing the interaction between the faster storm-scale instabilities and slower large-scale environment is a big challenge in constructing an ensemble breeding system for TC forecasts. Hence, this section presented a new TC breeding approach that could help improve this challenge.

2.2 TC breeding method

Though the first breeding method presented by TK93 could capture the trend of fastest-growing during a finite time window, the real world TCs have a finite life cycle. Due to the high-resolution regional modeling required large computation, TC predicting models are typically spark off only when their TCs are already first reported in the warning centers, because it is a challenge to conserve a continuous ensemble of breeding cycles with taking much computational capacity for a long time. Therefore, TK93's breeding scheme could not instantaneously acquire directions of most unstable modes during the earliest cycles. Moreover, utilizing only single re-scaling factor for both storm inner-core region and ambient environment with distinguished spatio-temporal scales does not enclose all mesoscale unstable nodes associated with TC vortex dynamics for which perturbations at different spatial–temporal scales grow at different rates [59]. Hence, it is necessary to change the rescaling factors following both the flow and the scales of instabilities [13].

Since our TC-breeding approach is focus on characterizing not only the storm-scale but also the large-scale unstable modes and their mutual interaction, there are two different scaling factors for these scale modes separately.

In the TK93's breeding extended design for TC predicting (hereafter known as the TC-breeding method or TCB), steps to make the TC-bred seeds as follows:

Step 1. Remove the GFS original vortex and insert a bogus vortex into the GFS initial condition to obtain a new first guess $\boldsymbol{x}^a$. In which, the bogus vortex is dynamical constructed based on the observed minimum sea-level pressure and maximum surface wind, using the Australian Bureau of Meteorology's Tropical Cyclone Limited Area Prediction System (TC-LAPS) package. This step is essential due to the weaknesses of the original GFS vortex in coarse resolution;

Step 2: adding and subtract a bred seeds d_i (i = 1,2, ... ,6), then we have 6 first guess $x1^a_i = x^a + d_i$ (positive sector) and $x2^a_i = x^a - d_i$ (negative sector)

Step 3: Run 6-hour lead time forecasts for both positive and negative sectors

Step 4: Separate 6-h forecasts (operators $\mathbf{S^m}$ and $\mathbf{S^v}$) of positive sector ($x1^f_i$) and negative sector ($x2^f_i$) from the previous breeding ensemble forecasts into an environmental component $\mathbf{S^m x1^f_i}$ and $\mathbf{S^m}$ $\mathbf{x2^f_i}$) and a vortex component ($\mathbf{S^v x1^f_i}$ and $\mathbf{S^v}$ $\mathbf{x2^f_i}$, **Figure 4a**).

Step 5: Find difference (operator H, **Figure 4a**) of each set of bred vector pairs (or seeds) from previous 6-h cycles to obtain environmental bred vectors $\boldsymbol{m}_i = \boldsymbol{S}^m\left(\boldsymbol{p}_i^f - \boldsymbol{n}_i^f\right)$ and the TC bred vectors $\boldsymbol{v}_i = \boldsymbol{S}^v\left(\boldsymbol{p}_i^f - \boldsymbol{n}_i^f\right)$;

Step 6: normalize the environmental bred vectors (by using a normalizing operator $\mathbf{C_m}$) to obtain a new set of normalized bred vector $C_m m_{i.,}$ then use an orthogonal operator T to obtain an orthogonal set of environmental bred vectors $\mathbf{TC_m m_i}$. Here, the environmental re-scaling operator $\mathbf{C^m}$ acting on a vector **v** is defined as:

$$\boldsymbol{C}^m\mathbf{v} \equiv \Lambda\frac{\mathbf{v}}{||\mathbf{v}||}, \tag{7}$$

With the scaling factor for the environmental perturbations given by

$$\Lambda = \left[\frac{1}{2\Gamma}\int_D\int_z\left[U'^2 + V'^2 + \frac{C_p}{T}T'^2\right]dzdS\right]^{\frac{1}{2}}, \tag{8}$$

And the norm ||.|| taken to be the energy norm as follows:

$$||\mathbf{v}||^2 = \left[\frac{1}{2\Gamma}\int_D\int_z\left[u'^2 + v'^2 + \frac{C_p}{T_0}T'^2\right]dzdS\right]^{\frac{1}{2}},$$

where Γ is the normalized factor proportional to the model domain volume, C_p = 1006 J kg^{-1} K^{-1}; T0 = 300 K), D is the model domain area after the model vortex was filtered, $U' = V' = 1.8$ ms^{-1}, and $T' = 0.7$K. These values are established in the study of Saito et al. [13], which are also consistent with the previous estimation by Wang and Bishop [60]; Repeat step 6 for the TC bred vectors with operator C_x to obtain a set of orthogonal TC bred vectors $\mathbf{TC_x}v_i$.

Step 7. Make a new pairs of breeding members by adding/subtracting the environmental and TC bred components into the analysis $\mathbf{x^a}$, i.e.,
$\mathbf{x1^a}_i = \mathbf{x^a} + \mathbf{TC_m m_i} + \mathbf{TC_x v_i}$ and $\mathbf{x2^a}_i = \mathbf{x^a} - \mathbf{TC_m m_i} - \mathbf{TC_x v_i}$.

Step 8: Run 6-hour lead time forecasts of all positive and negative pair to serve as the first guess for the next analysis cycle;

Step 9. Repeat step 1–9 for the next analysis cycle;

Noted that the above steps are taken only for the warm-start mode in which the ensemble breeding forecasts in the analysis procedure has been available since the previous. For the cold start cycle at which the "INVEST" information for a tropical depression is first issued, it is apparent that the bred vectors are unknown yet, therefore the ensemble initialization requires a different procedure.

One can do the cold-start in countless ways, for example using a random Gaussian noise with a prescribed error distribution, or directly use of the global GFS ensemble forecasts. For simplicity, the approach uses the 6-h difference from previous GFS short-range forecasts for all of the cold-start ensembles. This approach, known as lagged-averaged forecasts (LAF) from Kalnay [58], can quickly capture the most unstable modes in the model, thus allowing the breeding ensemble to speed up the dynamically representation to the environment. The combination of these short-range forecasts can generate a predefined number of seeds from which the breeding ensemble can be obtained. Consider, for example, a configuration of the breeding ensemble experiments requires a total of six bred vectors. Those bred vectors are initialized by taking six 6-h differences of the previous -36 h, −24 h, −18 h, −12 h, and -6 h forecasts that are all taken from the cold start ensemble (**Figure 4b**). The control forecast preprocessed directly from the GFS analysis then adds/subtracts the given bred vectors to create an ensemble of total 13 members for subsequent ensemble forecasts.

2.2.1 Example 2

The TC Breeding method has been implemented the Regional Atmospheric Modeling System (RAMS, version 6.0) model to forecast the TC track in the WPAC basin. In this study, the model domain is a region limited by 5°S–35°N and 100–150°E. This domain is sufficiently large to cover most of the tropical cyclone that formed in the WPAC basin and part of the Tibetan plateau that affects the large-scale steering flow of the TC tracks in the WPAC basin. The model integration time is 60s, and the experimental maximum lead times were up to 5 days (120 h). The convection parameterization schemes used among all experiments included a Kuo scheme, a Kain–Fristch scheme (original) and the new Kain–Fristch scheme (modified version). Initial data for model input were taken from the National Center for Environmental Prediction (NCEP) Global Forecast System (GFS) operational forecast with resolution of 1° × 1°. A set of 14 tropical cyclones between 2009 and 2011 in the WPAC basin were chosen for testing the TCB method (**Table 1**).

A series of 120 h forecasts for all storms in **Table 1** were conducted, using the aforementioned TC-breeding technique. The retrospective experiments include six positive/negative pairs and a control forecast (total 13 members). Here, the control forecasts are just the integrated results from the RAMS model with initial conditions where the original GFS forecasts adding a bogus vortex to make sure the model storm intensity was equivalent to the reality. The experiments used the default mode of the TC-LAPS package in which the constructed bogus vortex that had the horizontal resolution of 1° × 1°, and the isobaric vertical coordinates with 26

No	Name	Start date	End date
2009			
1	CHANHOM	18z02052009	00z09052009
2	LINFA	06z17062009	12z22062009
3	GONI	00z30072009	12z09082009
4	MUJIGAE	12z08092009	00z12092009
5	KETSANA	00z25092009	06z30092009
6	PARMA	18z28092009	18z28092009
7	MIRINAE	18z26102009	12z02112009
2010			
8	CONSON	18z11072010	18z17072010
9	CHANTHU	00z18072010	06z23072010
2011			
10	HAIMA	00z19062011	18z24062011
11	NOCKTEN	06z25072011	15z30072011
12	NESAT	00z24092011	12z30092011
13	NALGAE	00z28092011	06z05102011
14	WASHI	00z15122011	18z19122011

Table 1.
List of storms between 2009 and 2011 in the WPAC basin used in this study.

pressure levels as 1000, 975, 950, 925, 850, 800, 750, 700, 650, 600, 550, 500, 450, 400, 350, 300, 250, 200, 150, 100, 70, 50, 30, 20, and 10 h Pa. The variables to characterize the bogus vortex included sea level pressure (P), horizontal wind components (U, V), temperature (T), geopotential height (H), and relative humidity (RH). The cycles of all breeding ensemble were every 12 h, using the TCB method described above. Besides, for convenience, the domain of storm-scale perturbation was also fixed with enclosing area of 1000 km × 1000 km, centered at the vortex location. Three perturbed state variables in the model at the breeding cycles included the horizontal winds and potential temperature at all pressure levels. These cycles (12-h interval) are suitable enough to capture both the fast-growing weather signals at the micro- to meso- scale and the slower baroclinic modes at larger scales.

Results indicated that TCB method helps reduce the track errors. The improvement is approximately 10% reduction in the track forecast errors at the 4- to 5- day lead times as compared to deterministic forecasts integrated from GFS derived-initial conditions. While the improvement is not significant at shorter lead time (1–3 lead times, **Figure 5**).

Besides, the major difference between this TCB method and the original approach of TK93 is the dissimilarity in treatment of perturbations between large-scale environments and storm-scale inner-core, which are then orthogonalized in different manners. For the environmental perturbations in all experiments, a volume limited by [100–150°E] × [5°S–35°N] × [1000–10 h Pa] is chosen. In facts, the domain size does not have significant impact on the magnitude of EBVs, when it is important for the TBVs in some aspects. That is because storms do not always have a fixed size, thus the use of a predefined domain with a constant radius of 1000 km may not fully characterize the storm-scale TC-like vortex. One can design a suitable

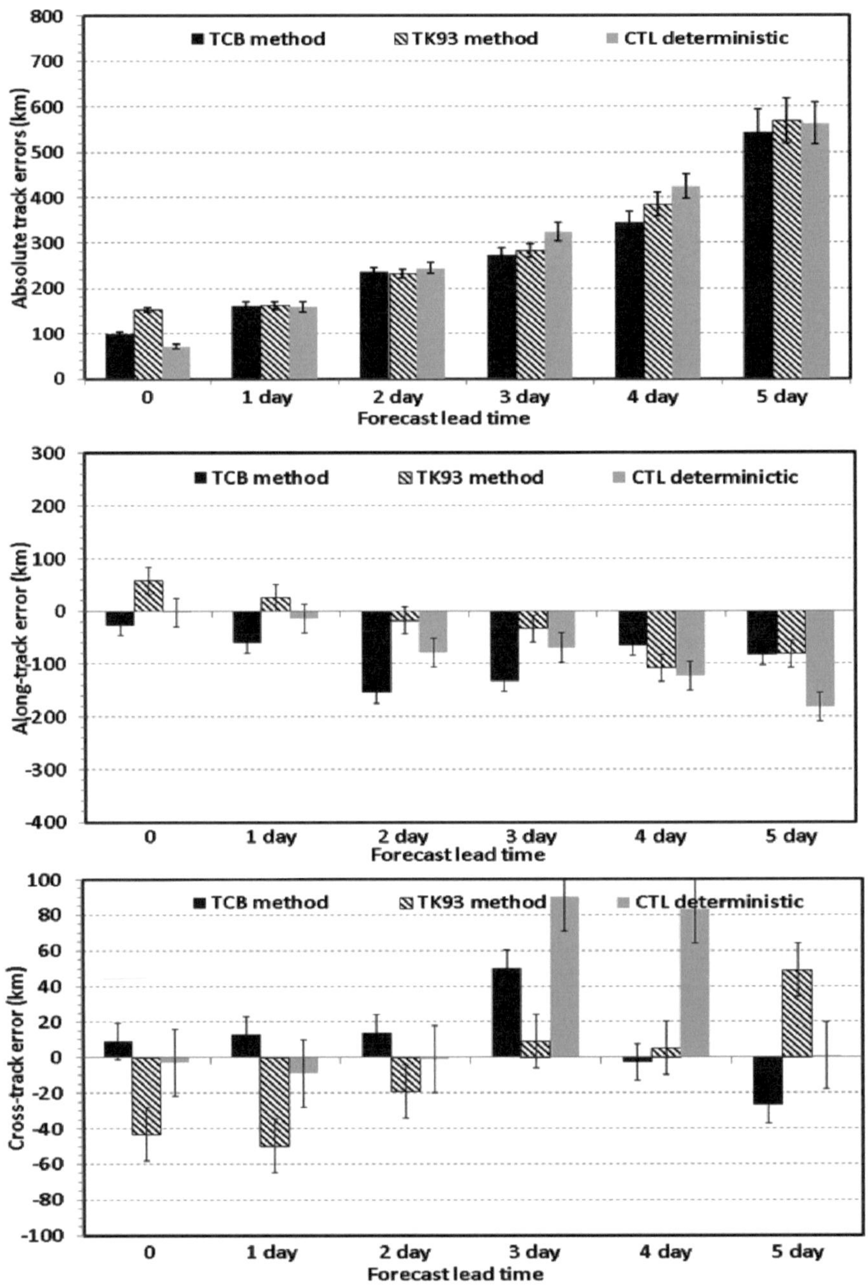

Figure 5.
Track forecast distance errors between the TK93's original breeding ensemble (striped column) forecasts and the deterministic control forecasts (gray columns) for 2009–2011 seasons using the RAMS model [47].

adaptive storm domain to optimize the effectiveness of the TCB method. However, with a coarse resolution of 30 km, the adaptive approach cannot capture the true detailed TC inner-core structure. For more simplicity of the experiment design in this study, the filtering domain has a fixed horizontal radius of 1000 km in all experiments with a warning that this constant size could be a caveat for very broad TCs. It should be noted also that the control analysis would add or subtract the bred vectors, and a potential drift of the control run from the actual states may shift the entire ensemble further from the truth after several cycles. However, with the

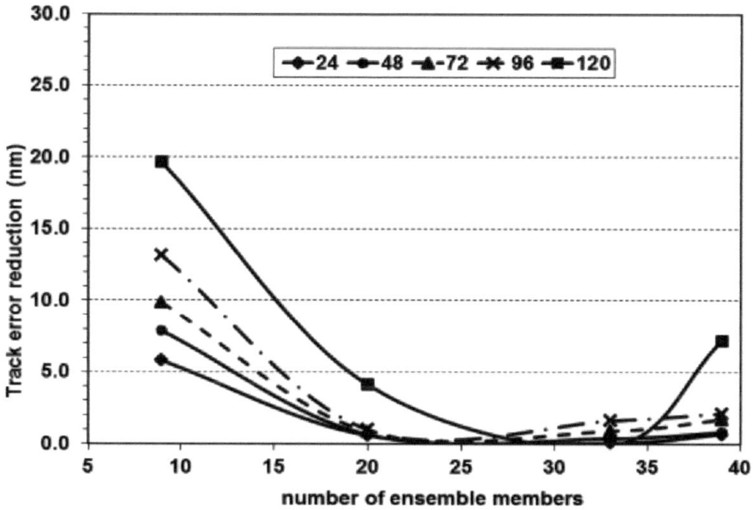

Figure 6.
Rate of track forecast error base on the number of the ensemble members for forecast ranges: The 24-h (diamond), 48-h (circle), 72-h (triangle), 96-h (times), and 120-h (square). The reducing rate is determined as the difference of the track errors when adding newly member to the system at each lead time [56].

integration lead time of only 12 h at each cycle from the control forecast using the GFS forecasts, such a drift is not a big issue and the ensemble thus always maintains their close trajectory to the truth at every initial time.

Sensitivity experiments showed that the best results with ~30 ensemble members are adequate to construct a TCB technique. By gradually increasing the number of ensemble members, the rate of reducing track error per newly added member becomes saturated after reaching the number of 30 ensemble members (**Figure 6**). This saturation of the track errors could link to the maximum information that the orthogonalization of the bred vectors to be obtained after the system reaches its noise level. Otherwise, adding more ensemble member could provide no further benefit to the system, it could even slow down the computation. It should be noted that the 30-km resolution of all ensemble experiments does not fully verify the necessity of separating the treatments for the storm-scale bred vectors and the large-scale bred vectors in distinguished manners. Theoretically, one could design the experiments with higher resolution to further assess the sensitivity of the breeding ensemble technique for more precise experiments, but this would require a large amount of computational and storage resources beyond our current capability. Although this minor problem about resolution, the overall track forecast improvement with the TCB approach suggests that this approach could somehow shed light on ensemble TC track forecast, especially under the circumstances where the observational information is not enough to execute more complex data assimilation steps in real-time forecasting systems.

3. Conclusions

This chapter has presented several techniques to improve the predictive quality of tropical cyclone formation and trajectory. For the forecast of TCs formation, the LETKF algorithm and its implementation in the WRF model and the Vortex tracking method have been introduced. Results in example 1 show that due to a better approach in capturing the real world monsoon trough by assimilating augmented observations available during the early stages of TC Wutip, the WRF-LETKF model had provided better forecasts about the formation location and timing of typhoon Wutip in comparison to the forecasts that used initial conditions directly from GFS global model. Besides, the results from this study also show the CIMSS-AMV data played a vital role in improving the information of the large-scale environment required for TC formation that one should consider for real-time TC forecasts. For the tropical cyclone track forecasts, a breeding ensemble technique is introduced. This technique is developed based on the original breeding method (TK93). Experiments with 14 TCs (**Table 1**) in **example** showed a promising reduction of track forecast errors by using the TCB technique, especially at 4–5 days forecast range.

However, both the track forecasts by TCB method and the control forecasts are similar in the patterns of cross- and along track forecast errors. This indicated that model inherent errors also are a significant contributor to the track forecast errors that the TCB method is unable to eliminate. Sensitivity experiments of adding gradually each ensemble members exhibit further that the increasing number of members could reduce the track forecast errors, but reduction rate saturates when the number reaches 30 dues to the inefficiency of the TCB method in orthogonalizing bred vectors. However, while the TCB method cannot eliminate model inherent errors related to inadequate representation of sub-grid scales when using only parameterizations of physical processes in the RAMS model or the inefficient model resolution, this method could somehow optimize the use of the breeding ensemble technique for tropical cyclone track forecasts in real-time forecasting systems which do not require high computational resources.

Author details

Cong Thanh, Dao Nguyen Quynh Hoa and Tran Tan Tien*
VNU University of Science, Hanoi, Vietnam

*Address all correspondence to: tientt49@gmail.com

References

[1] Chen J., Lin S., Zhou L., Chen X., Rees S., Bender M., and Morin M.: Evaluation of tropical cyclone forecasts in the next generation global prediction system. Mon. Wea. Rev.. 2019;147; 3409–3428. DOI: https://doi.org/10.1175/MWR-D-18-0227.1

[2] Zhao M., Held I. M., Lin S.-J., and Vecchi G. A.: Simulations of global hurricane climatology, interannual variability, and response to global warming using a50-km resolution GCM. J. Climate. 2009;22;6653–6678. DOI: https://doi.org/10.1175/2009JCLI3049.1.

[3] Chan J. C. L., and Kwok R. H.: Tropical cyclone genesis in a global numerical weather prediction model. Mon. Wea. Rev.. 1999;127;611–624. DOI: https://doi.org/10.1175/1520-0493(1999)127<0611:TCGIAG>2.0.CO;2

[4] Cheung K. K., and Elsberry R. L.: Tropical cyclone formations over the western North Pacific in the Navy Operational Global Atmospheric Prediction System forecasts. Wea. Forecasting. 2002;17;800–820. DOI: https://doi.org/10.1175/1520-0434(2002)017<0800:TCFOTW>2.0.CO;2.

[5] Halperin D. J., Fuelberg H. E., Hart R. E., Cossuth J. H., Sura P., and Pasch R. J.: An evaluation of tropical cyclone genesis forecasts from global numerical models. Wea. Forecasting. 2013;28; 1423–1445. DOI: https://doi.org/10.1175/WAF-D-13-00008.1.

[6] Halperin D. J., Fuelberg H. E., Hart R. E., Cossuth J. H.: Verification of tropical cyclone genesis forecasts from global numerical models: Comparisons between the North Atlantic and eastern North Pacific basins. Wea. Forecasting. 2016;31; 947–955. DOI: https://doi.org/10.1175/WAF-D-15-0157.1.

[7] Oouchi K., Yoshimura J., Yoshimura H., Mizuta R., Kusunoki S., and Noda A.: Tropical cyclone climatology in a globalwarming climate as simulated in a 20 km-mesh global atmospheric model: Frequency and wind intensity analyses. J. Meteor. Soc. Japan. 2006;84; 259–276. DOI: https://doi.org/10.2151/jmsj.84.259

[8] Szunyogh I., Kostelich E. J., Gyarmati G., Kalnay E., Hunt B. R., Ott E., Satterfield E., and Yorke J. A.: A local ensemble transform Kalman filter data assimilation system for the NCEP global model. Tellus. 2008;60A;113–130. DOI: https://doi.org/10.1111/j.1600-0870.2007.00274.x

[9] Zhou L., Lin S.-J., Chen J.-H., Harris L. M., Chen X., and Rees S.: Toward convective-scale prediction within the Next Generation Global Prediction System. Bull. Amer. Meteor. Soc.. 2019; 100;1225–1243. DOI: https://doi.org/10.1175/BAMS-D-17-0246.1.

[10] Aberson S. D., Aksoy A., Sellwood K. J., Vukicevic T., and Zhang X.: Assimilation of high-resolution tropical cyclone observations with an ensemble Kalman filter using HEDAS: Evaluation of 2008–2011 HWRF forecasts. Mon. Wea. Rev.. 2015;143; 511–523. DOI: https://doi.org/10.1175/MWR-D-14-00138.1.

[11] Holt C., Szunyogh I., Gyarmati G., Leidner S. M., and Hoffman R. N.: Assimilation of tropical cyclone observations: Improving the assimilation of TCVitals, scatterometer winds, and dropwindsonde observations. Mon. Wea. Rev.. 2015;143; 3956– 3980. DOI: https://doi.org/10.1175/MWR-D-14-00158.1.

[12] Kieu C. Q., Minh P. T., and Mai H. T.: An application of the multi-physics ensemble Kalman filter to typhoon forecast. Pure Appl. Geophys.. 2013;171; 1473–1497. DOI: https://doi.org/10.1007/s00024-013-0681-y

[13] Saito K., and Coauthors: The WWRP Beijing Olympic 2008 RD Project. Meeting on the Study of data assimilation and evaluation of forecast reliabilities for dynamical prediction of heavy rainfall. Meteorological Research Institute. Mar 2008, Japan.

[14] Tong M., and Coauthors: Impact of assimilating aircraft reconnaissance observations on tropical cyclone initialization and prediction using operational HWRF and GSI ensemble–variational hybrid data assimilation. Mon. Wea. Rev.. 2018;146;4155–4177. DOI: https://doi.org/10.1175/MWR-D-17-0380.1

[15] Zhang F., and Weng Y.: Predicting hurricane intensity and associated hazards: A five-year real-time forecast experiment with assimilation of airborne Doppler radar observations. Bull. Amer. Meteor. Soc.. 2015;96;25–33. DOI: https://doi.org/10.1175/BAMS-D-13-00231.1

[16] Zhang F., Weng Y., Gamache J. F., and Marks F. D.: Performance of convection-permitting hurricane initialization and prediction during 2008–2010 with ensemble data assimilation of innercore airborne Doppler radar observations. Geophys. Res. Lett.. 2011;38;L15810. DOI: https://doi.org/10.1029/2011GL048469.

[17] Du T. D., Ngo-Duc T., and Kieu C.: Initializing the WRF Model with tropical cyclone real-time reports using the ensemble Kalman filter algorithm. Pure Appl. Geophys.. 2017;174;2803–2825. DOI: https://doi.org/10.1007/s00024-017-1568-0.

[18] Pu Z., Zhang S., Tong M., and Tallapragada V.: Influence of the self-consistent regional ensemble background error covariance on hurricane inner-core data assimilation with the GSI-based hybrid system for HWRF. J. Atmos. Sci.. 2016;73;4911–4925. DOI: https://doi.org/10.1175/JAS-D-16-0017.1.

[19] Snyder C., and Zhang F.: Assimilation of simulated Doppler radar observations with an ensemble Kalman filter. Mon. Wea. Rev.. 2003;131;1663–1677. DOI: https://doi.org/10.1175//2555.1

[20] Ott E., and Coauthors: A local ensemble Kalman filter for atmospheric data assimilation. Tellus. 2004;56A;415–428. DOI: https://doi.org/10.3402/tellusa.v56i5.14462

[21] Hunt B. R., Kostelich E. J., and Szunyogh I.: Efficient data assimilation for spatiotemporal chaos: A local ensemble transform Kalman filter. Physica D.. 2007;230;112–126. DOI: https://doi.org/10.1016/j.physd.2006.11.008.

[22] Holmlund K., Velden C., and Rohn M.: Enhanced automated quality control applied to high-density satellite-derived winds. Mon. Wea. Rev.. 2001;129;517–529. DOI: https://doi.org/10.1175/1520-0493(2001)129<0517:EAQCAT>2.0.CO;2

[23] Kang J.-S., Kalnay E., Liu J., Fung I., Miyoshi T., and Ide K.: "Variable localization" in an ensemble Kalman filter: Application to the carbon cycle data assimilation. J. Geophys. Res.. 2011;116;D09110. DOI: https://doi.org/10.1029/2010JD014673

[24] Kieu C. Q., Nguyen M. T., Hoang T. M., and Ngo-Duc T.: Sensitivity of the track and intensity forecasts of Typhoon Megi (2010) to satellite-derived atmospheric motion vectors with the ensemble Kalman filter. J. Atmos. Oceanic Technol.. 2012;29;1794–1810. DOI: https://doi.org/10.1175/JTECH-D-12-00020.1.

[25] Miyoshi T., and Kunii M.: The local ensemble transform Kalman filter with the weather research and forecasting

model: Experiments with real observations. Pure Appl. Geophys.. 2012;169;321–333. DOI: https://doi.org/10.1007/s00024-011-0373-4.

[26] Cheung K. K. W., Chan J. C. L.: Ensemble Forecasting of Tropical Cyclone Motion Using a Barotropic Model. Part I: Perturbations of the Environment. Mon. Wea. Rev.. 1999; 127;1229–1243. DOI: https://doi.org/10.1175/1520-0493(1999)127<1229:EFOTCM>2.0.CO;2.

[27] Li H., Kalnay E., Miyoshi T., and Danforth C. M.: Accounting for model errors in ensemble data assimilation. Mon. Wea. Rev.. 2009;137;3407–3419.

[28] Miyoshi T.: The Gaussian Approach to Adaptive Covariance Inflation and Its Implementation with the Local Ensemble Transform Kalman Filter. Mon. Wea. Rev.. 2011;139;1519–1535.

[29] Bister M., and Emanuel K. A., 1997: The genesis of Hurricane Guillermo: TEXMEX analyses and a modeling study. Mon. Wea. Rev.. 1997;125;2662–2682. DOI: https://doi.org/10.1175/1520-0493(1997)125<2662:TGOHGT>2.0.CO;2

[30] Wen D., Li Y., Zhang D., Xue L., and Wei N.: A statistical analysis of tropical upper-tropospheric trough cells over the western North Pacific during 2006–15. J. Appl. Meteor. Climatol.. 2018;57;2469–2483. DOI: https://doi.org/10.1175/JAMC-D-18-0003.1

[31] Le Marshall J., Rea A., Leslie L., Seecamp R., and Dunn M.: Error characterisation of atmospheric motion vectors. Aust. Meteor. Mag.. 2004;53; 123–131.

[32] Li J., Li J., Velden C., Wang P., Schmit T. J., and Sippel J.: Impact of rapid-scan-based dynamical information from GOES-16 on HWRF hurricane forecasts. J. Geophys. Res. Atmos.. 2020;125;e2019JD031647. DOI: https://doi.org/10.1029/2019JD031647

[33] Velden C. S., Hayden C. M., Menzel W. P., Franklin J. L., and Lynch J. S.: The impact of satellite-derived winds on numerical hurricane track forecasting. Wea. Forecasting. 1992;7;107–118. DOI: https://doi.org/10.1175/1520-0434(1992)007<0107:TIOSDW>2.0.CO;2

[34] Velden C. S., Hayden C., Nieman S., Menzel W., Wanzong S., and Goerss J.: Upper-tropospheric winds derived from geostationary satellite water vapor observations. Bull. Amer. Meteor. Soc.. 1997;78;173–195. DOI: https://doi.org/10.1175/1520-0477(1997)078<0173:UTWDFG>2.0.CO;2

[35] Tien T. T., Dao N. Q. H., C. Thanh, Kieu C. Q.: Assessing the Impacts of Augmented Observations on the Forecast of Typhoon Wutip's (2013) Formation Using the Ensemble Kalman Filter. Weather and Forecasting. 2020; 35;1483–1503.

[36] Cecelski, S. F. and D.-L. Zhang (2013). "Genesis of Hurricane Julia (2010) within an African easterly wave: Low-level vortices and upper-level warming." J. Atmos. Sci. **70**: 3799-3817.

[37] Cecelski, S. F., Zhang, D., & Miyoshi, T. (2014). Genesis of Hurricane Julia (2010) within an African Easterly Wave: Developing and Nondeveloping Members from WRF–LETKF Ensemble Forecasts, *Journal of the Atmospheric Sciences*, *71*(7), 2763-2781

[38] Kishimoto K.: JMA's five-day tropical cyclone track forecast. Technical Review No. 12 of the RSMC Tokyo – Typhoon Center. Available online at http://www.jma.go.jp/jma/jmaeng/jma-center/rsmc-hp-pub-eg/techrev/text12-2.pdf. 2010.

[39] Zhang Z., and Krishnamurti T. N.: "Ensemble forecasting of hurricane

tracks." Bull Amer. Meteor. Soc. 1997;78; 2785-2795.

[40] Elsberry R. L., and Carr III L. E.: Consensus of dynamical tropical cyclone track forecasts—Error versus spread. Mon. Wea. Rev.. 2000;128;4131–4138. DOI: https://doi.org/10.1175/1520-0493(2000)129<4131:CODTCT>2.0.CO;2

[41] Kehoe R. M., Boothe M. A., Elsberry R. L.: Dynamical Tropical Cyclone 96- and 120-h Track Forecast Errors in the Western North Pacific. Wea. Forecasting. 2007;22;520–538. DOI: https://doi.org/10.1175/WAF1002.1

[42] Tien T. T., C. Thanh, Van H. T., Kieu C. Q.: Two-dimensional retrieval of typhoon tracks from an ensemble of multi-model outputs. Wea. Forecasting. 2012;27;451–461.

[43] Toth Z., Kalnay E.: "Ensemble forecasting at NMC." Amer. Meteor. Soc.. 1993;74;2317-2330.

[44] Toth Z., and Kalnay E.: Ensemble forecasting at NCEP and the breeding method. Mon. Wea. Rev.. 1997;125; 3297–3319.

[45] Molteni F., Buizza R., Palmer T. N., and Petroliagis T.: "The ECMWF ensemble prediction system: Methodology and validation. ."Quart. J. Roy. Meteor. Soc.. 1996;122;73-119.

[46] Palmer T. N., Molteni F., Mureau R., Buizza R., Chapelet P., Tribbia J.: Ensemble prediction. In Proc. of the ECMWF Seminar on Validation of Models over Europe. Vol. 1 (ECMWF, Shinfield Park, Reading RG2 9AX, UK). 1993. 285 pp.

[47] C. Thanh, Tien T. T., and Chanh K. Q.: Application of breeding ensemble to tropical cyclone track forecasts using the regional atmospheric modeling system (rams) model. Applied Mathematical Modelling. 2016;40(19–20);8309–8325. DOI: https://doi.org/10.1016/j.apm.2016.04.010

[48] Magnusson L, Källën E., and Nycander J.: "Initial state perturbations in ensemble forecasting." Nonlin. Processes Geophys.. 2008;15;751–759.

[49] Corazza M., Kalnay E., Patil D. J., Ott E., Yorke J., Szunyogh I., Cai M.: Use of the breeding technique in the estimation of the background error covariance matrix for a quasigeostrophic model, in: AMS Symposium on Observations, Data Assimilation and Probabilistic Prediction, Orlando, Florida. 2002. pp. 154-157.

[50] Trevisan A., and Pancotti F.: Periodic orbits, Lyapunov vectors and singular vectors in the Lorenz system. J. Atmos. Sci.. 1998;55;390-398.

[51] Annan J. D.: On the orthogonality of bred vectors. Mon. Weather Rev., 2004; 843-849.

[52] Yang S.-C., Kalnay E., Cai M., Rienecker M., Yuan G., Toth Z.: ENSO Bred Vectors in Coupled Ocean–Atmosphere General Circulation Models. J. Climate. 2006;19;1422–1436.

[53] Legras B., and Vautard R.: A guide to Lyapunov vectors. Proceedings of the ECMWF Seminar on Predictability. September 4–8, 1995, Reading, England, Vol 1, ECMWF, Shinfield Park, Reading, UK. 1996. pp. 143-156.

[54] Vannitsem S., and Nicolis C.: Lyapunov vectors and error growth patternsin a T21L3 quasigeostrophic model. J. Atmos. Sci.. 1997;54;347-361.

[55] Black M. L., and Willoughby H. E.: The concentric eyewall cycle of Hurricane Gilbert. Mon. Wea. Rev.. 1992;120;947–957. DOI: https://doi.org/10.1175/1520-0493(1992)120<0947:TCECOH>2.0.CO;2

[56] Chen Y., and Yau M. K.: Spiral bands in a simulated hurricane. Part I: Vortex Rossby wave verification. J.

Atmos. Sci.. 2001;58;2128–2145. DOI: https://doi.org/10.1175/1520-0469 (2001)058<2128:SBIASH>2.0.CO;2

[57] Guinn T. A., and Schubert W. H.: Hurricane spiral bands.J. Atmos. Sci.. 1993;50;3380–3403.

[58] Kalnay E.: "Atmospheric modeling, data assimilation and predictability", Cambridge University Press. 2003;512p.

[59] Lorenz E. N.: "The predictability of a flow which possesses many scales of motion." Tellus. 1969;21;289-307.

[60] Wang X., Bishop C. H.: A Comparison of Breeding and Ensemble Transform Kalman Filter Ensemble Forecast Schemes. J. Atmos. Sci.. 2003; 60;1140–1158.

Chapter 4

Prediction of Relative Humidity in a High Elevated Basin of Western Karakoram by Using Different Machine Learning Models

Muhammad Adnan, Rana Muhammad Adnan, Shiyin Liu, Muhammad Saifullah, Yasir Latif and Mudassar Iqbal

Abstract

Accurate and reliable prediction of relative humidity is of great importance in all fields concerning global climate change. The current study has employed Multivariate Adaptive Regression Spline (MARS) and M5 Tree (M5T) models to predict the relative humidity in the Hunza River basin, Pakistan. Both the models provided the best prediction for the input scenario S6 (RHt-1, RHt-2, RHt-3, Tt-1, Tt-2, Tt-3). The statistical analysis displayed that the MARS model provided a better prediction of relative humidity as compared to M5T at all meteorological stations, especially, at Ziarat followed by Khunjerab and Naltar. The values of root mean square error (RMSE), mean absolute error (MAE), and coefficient of determination (R^2) were (5.98%, 5.43%, and 0.808) for Khunjerab; (6.58%, 5.08%, and 0.806) for Naltar; and (5.86%, 4.97%, 0.815) for Ziarat during the testing of MARS model whereas, the values were (6.14%, 5.56%, and 0.772) for Khunjerab; (6.19%, 5.58% and 0.762) for Naltar and (6.08%, 5.46%, 0.783) for Ziarat during the testing of M5T model. Both the models performed slightly better in training as compared to the testing stage. The current study encourages future research to be conducted at high altitude basins for the prediction of other meteorological variables using machine learning tools.

Keywords: relative humidity, MARS, M5T, Hunza, machine learning

1. Introduction

The relative humidity is defined as the amount of water vapor in the air in comparison with the full saturation [1, 2]. Being the important indicator of precipitation forecasting, its prediction plays a significant part in improving the accuracy of weather forecasting [3]. The relative humidity changes with respect to change in saturated vapor pressure which further depends on wind speed, solar radiation, pressure, temperature, and moisture content in the air [1]. The relative humidity is a function of temperature and is regarded as a sensitive parameter in the field of science [4]. Relative humidity plays a vital role in plant growth, agricultural and industrial production and in the prevention and control of air pollution [5];

IntechOpen

economic stability of a region, water systems and also in managing renewable and solar energy systems [1, 6], weather and climate [7, 8]. Moreover, it has also an impact on ozone concentration and adaptive thermal comfort [9]. Keeping in view the importance of relative humidity, the research on its prediction is increasingly important [7].

The relative humidity is an important aspect of the hydrological phase [8] and has a role in alpine hydrology, especially, in a cold and dry climate; any change in temperature and humidity causes larger variations in the ablation of glaciers [10]. The warm environment glaciers are subjected to be influenced more by the change in relative humidity. Few other studies e.g. [11–13] also observed that tropical glaciers are sensitive to subtle changes in relative humidity, precipitation, and cloudiness. Relative humidity and clouds play an important role in the energy balance of glaciers by controlling the number of outgoing longwave radiation. Moreover, relative humidity and wind speed influence the turbulent latent heat flux which supplies all energy for sublimation and thus they indirectly control the equilibrium line altitude (ELA) [14]. Another study conducted by [15] observed that relative humidity has an effect on evaporation and there is an inverse relation between them. Evaporation further controls the water balance of closed lakes in hilly areas and evapotranspiration, especially, in irrigated agricultural areas.

Regardless of relative humidity is an important component of hydrology, meteorology, and climate, only a few studies are available for its prediction. A study conducted by [1] used artificial neural networks (ANNs) and genetic expression programming (GEP) models for the prediction of relative humidity as a function of three meteorological variables: wind speed, temperature, and pressure in two Californian gauging stations. They observed that both the models can successfully predict one-year relative humidity data into the future. Another study done by [5] predicted relative humidity by establishing time series models such as Extreme Gradient Boosting (XGBoost), Seasonal Auto-Regressive Integrated Moving Average (SARIMA), and Holt-Winters (HW). The XGBoost was found more accurate because of its robust capability to resist a fitting. The study conducted by [3] found that the performance of an autoregressive integrated moving average (ARIMA) model is better than the Long Short-Term Memory (LSTM) Network for the prediction of relative humidity. On contrary, [8] observed that the LSTM network is capable of predicting complex univariate relative humidity time series with robust no-stationarity. However, Least Square Support Vector Machine (LSSVM) and Adaptive Network-Based Fuzzy Inference System (ANFIS) models were used by [2] for prediction of relative humidity in terms of dry bulb temperature and wet bulb depression and found satisfactory.

Another study conducted by [16] proposed four ANNs models to predict the relative humidity and temperature in a swine livestock warehouse located in Puerto Gaitan–Meta. They observed that the models used in the study are suitable for the prediction of humidity in barns not equipped with humidity sensors. However, [17] used an improved backpropagation (BP) neural network for the prediction of indoor relative humidity and temperature every 10 min and 6–72 hours in advance based on a cloud database in Chongqing, China. Both temperature and humidity predictions have a strong correlation with the observed data. Similarly, another study conducted by [18] used BP neural network for the prediction of one day ahead mean air temperature and relative humidity of greenhouse located in the sub-humid sub-tropical regions of India. The results displayed that the BP neural network model provided the best prediction for inside temperature and relative humidity. However, a study done by [19] used daily minimum air temperature (Tn) downscaled from INMCM4 general circulation model (GCM) to predict the relative humidity for climate change studies but relative humidity predictions were poor in few months especially in March, July, August, and October. Moreover, a study conducted by [20] proposed a Functional Link Neural Network (FLNN) which comprises of a single layer of tunable weight trained with the Modified Cuckoo Search algorithm (MCS) for prediction of daily temperature and relative humidity. It was observed that FNN when trained with MCS produced less prediction error. Further, an attempt has been made for the prediction of relative humidity and temperature at different locations inside tobacco dryer by [21] by using a fitting ANN model. Another study performed by [22] also used different ANN models to successfully forecast indoor relative humidity and temperature in the education building of Izmir, Turkey.

Formerly, no attempt has been made for the prediction of relative humidity in the alpine catchment where there is an issue of data scarcity. The current study is unique because it uses two machine learning models such as MARS and M5T to predict the relative humidity in the Hunza basin (glaciated basin), Pakistan. MARS model was selected because it requires a short training process and has the ability to model complex nonlinear processes deprived of strong model assumptions as compared to ANNs models [23, 24] whereas the M5T model was selected because of its small computation cost and ease in large data treatment as compared to support vector machine (SVM) and ANN [25, 26]. In previous studies, mostly these models were used for the prediction of runoff in poorly gauged basins. A study conducted by [27] suggested that the MARS method is capable of predicting short-term runoff forecast in mountainous watersheds whereas MARS was successfully used for the prediction of streamflows with inadequate data input in the mountainous catchment by [28]. Similarly, the M5T model was found useful in the prediction of streamflows of several tributaries by [29] and it was observed that predictions are good in rainless periods. Another study conducted by [30] found the M5T algorithm reliable in the prediction of streamflows. Several other studies also encouraged the researchers to use MARS and M5T models for the prediction of runoff e.g. [31–37]. Apart from runoff prediction, MARS and M5T models were also used for the prediction of evapotranspiration (ET) and Pan Evaporation (Ep). A study conducted by [38] compared the performance of M5T, MARS along with calibrated Hargreaves-Samani (CHS), MLP, and Stephens-Stewart (SS) models and observed that MARS performed better in the prediction of Ep. Another study conducted by [39] found that the M5T model outperformed compared to Ritchie Equation for the prediction of ET. Similarly, [40] successfully predicted reference evapotranspiration by using M5T and ANN models.

2. Study area

Hunza is a glaciated sub-catchment of the Upper Indus Basin (UIB) and is located in the western Karakoram Himalayan region of Pakistan (**Figure 1**). The basin lies within the extent of 74°02′–75°48′E and 35°54′–37°05′N and encompasses 13,671 km^2 of the catchment area.

The elevation of the basin ranged from 1391 to 7850 m. About 20% catchment area of the basin is covered by glaciers [41] and there are 110 glacial lakes in the basin [42]. It is the main tributary of the Indus Basin Irrigation System (IBIS) and it contributes about 12% of UIB streamflows upstream of Tarbela dam [43]. The climate of the Hunza basin is arid to semi-arid and is normally categorized by two seasons, October to March as winter and April to September as summer. The weather conditions vary within the basin. At low altitudes, weather is hot whereas at high altitudes winters are cold and there are extensive variations in temperature extremes [44]. The mean total annual precipitation varies with respect to altitude;

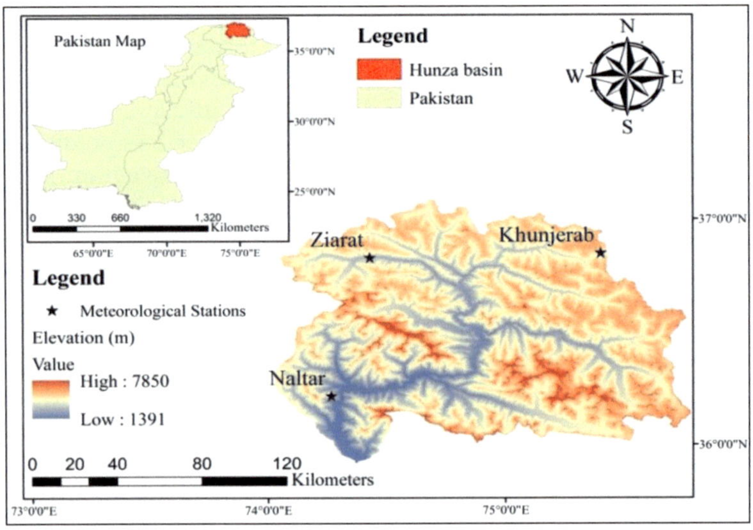

Figure 1.
Location map of the study area.

low altitude station such as Naltar (2858 m) receives more precipitation i.e. 660 mm as compared to high altitude station Khunjerab (4730 m) which receives 165 mm of precipitation. The meteorological station installed in between Naltar and Khunjerab (i.e. Ziarat, 3669 m) receives 292 mm of precipitation [45, 46].

The temporal variations in meteorological variables of Khunjerab station (using data of 1995–2009) are displayed in **Table 1**. **Table 1** shows that the maximum temperature varies between −11.1°C (January) to 11.6°C (July) whereas minimum temperature varies from −21.3°C (January) to 1.3°C (July). The maximum relative humidity in the basin varies from 59% (March) to 91% (August) while minimum relative humidity varies from 23% (March) to 52% (December). The daily solar radiation in the Hunza basin varies from 2563 (December) to 5148 (May) watt/m^2.

Month	Maximum Temperature (°C)	Minimum Temperature (°C)	Maximum Relative Humidity (%)	Minimum Relative Humidity (%)	Solar Radiation (watt/m^2)
January	−11.1	−21.3	62	30	2933
February	−11.0	−19.7	77	34	3500
March	−4.5	−16.9	59	23	4394
April	0.2	−9.7	78	30	4750
May	5.5	−4.6	81	26	5148
June	7.7	−1.3	87	44	5102
July	11.6	1.3	86	38	4858
August	10.5	−0.3	91	34	4711
September	4.6	−4.5	86	27	4227
October	0.8	−10.4	78	35	4003
November	−5.8	−16.1	68	39	3452
December	−11.0	−18.9	80	52	2563

Table 1.
Monthly average variations in meteorological variables of Khunjerab (1995–2012).

3. Material and methods

3.1 Topography

The Advanced Spaceborne Thermal Emission and Reflection Radiometer (ASTER), Global Digital Elevation Model (GDEM) was used to delineate the catchment boundary of the Hunza basin. The Hunza basin was delineated using ASTER GDEM v3 data in Arc GIS. The data was acquired from the website: https://lpdaac.usgs.gov/tools/data-pool/. The format of the downloaded tiles was Geo-TIFF and has the gridding resolution i.e. (30 m) and tile structure (1°x 1°).

3.2 Meteorological data

There are four meteorological stations in the Hunza River basin such as Hunza, Naltar, Khunjerab, and Ziarat (**Table 2**). The Hunza meteorological station was installed by the Pakistan Meteorological Department (PMD) and the record is available from 2007 to onward whereas the other three stations were installed and managed by Water and Power Development Authority (WAPDA) and the record is available from 1995 to onward. The current study has employed daily data of temperature, precipitation, solar radiation, and relative humidity of Ziarat, Naltar, and Khunjerab meteorological stations. The required data of the aforementioned stations were acquired from the Surface Water Hydrology Project of the Water and Power Development Authority (SWHP-WAPDA), Pakistan from 1995 to 2009 (**Table 2**).

3.3 Machine learning models

The current study has employed two machine learning models such as M5 Tree and MARS for the prediction of relative humidity at three meteorological stations of the Hunza basin. Their detailed description is given below:

3.3.1 M5 tree model

The M5T model was first introduced by [47]. Model trees simplify the theories of regression trees and there are constant values at their leaves [48]. M5T model is established in relation to a binary decision tree where linear regression functions are placed in the terminal node (leaf) and a relationship is developed between dependent and independent variables through it [49]. Model development involves two stages; the first stage involves in creation of a decision tree by using a split criterion

Meteorological Station	Latitude (*DD*)	Longitude (*DD*)	Elevation (m)	Data	Agency
Hunza	36.320	74.640	2374	P, T_{max}, T_{min}, RH, SR	PMD
Naltar	36.216	74.266	2858	—	SWHP-WAPDA
Khunjerab	36.850	75.400	4730	—	SWHP-WAPDA
Ziarat	36.830	74.430	3669	—	SWHP-WAPDA

Note: DD = Degree decimal; P= Precipitation; T_{max}= Maximum temperature; T_{min}= Minimum temperature; RH = Relative humidity; SR = Solar radiation.

Table 2.
List of meteorological stations in the Hunza basin.

whereas in the second stage overgrown tree is pruned for designing the model tree [25]. The splitting stage in the M5T model is composed of regression function at the leaves instead of class labels and continuous numerical attributes can be estimated through it [36]. The splitting criterion for the M5T model procedure is based on the standard deviation reduction (SDR) function achieved in every node. This criterion points out the error in that node and the minimum expected error is calculated by the model because of testing each attribute in that node [50, 51]. The SDR in the M5T model can be calculated by the following Equation [47]:

$$SDR = sd(M) - \sum \frac{|M_i|}{|M|} sd(M_i) \tag{1}$$

Where *SDR* specifies the standard deviation reduction and sd indicates standard deviation; *M* specifies a set of examples that reaches the node; whereas *Mi* signifies the subset of examples that have the ith outcome of the potential set.

Because of the splitting or branching process, data in child nodes (smaller nodes) have less SD than parent nodes (greater nodes). The division process often results in producing a large tree-like structure which causes overfitting and this issue can be resolved by pruning back the tree [52], for instance by substituting a subtheme with a leaf. Pruning the overgrown tree and substitution of subthemes with linear regression functions are performed in the second stage of model designing. This method of producing the model tree separates the parameter space into subspaces and builds in each of them a linear regression model.

3.3.2 MARS algorithm

MARS model was first developed by [53]. Its working procedure involved establishing a relationship among a set of input variables and the target-dependent that involve connections with less number of variables [54]. MARS produces flexible models to facilitate the solution space to be divided into several intervals of independent parameters whereas individual splines are fit to each interval [53]. This method is non-parametric and non-linear and it involves a forward-backward procedure to predict a continuous dependent parameter in high-dimensional data [55]. No assumptions have been made about the fundamental functional relationships between independent and dependent variables by the MARS model. In MARS, the splines are connected smoothly together to form piecewise curves which are also known as basis functions (BFs), and these form a flexible model which is capable of handling both linear and non-linear behavior [54]. Two stages are involved in setting up the MARS model which includes forward (constructing the model) and backward (a pruning procedure) stages. In the first stage (forward), to define a pair of BFs candidates, knots are placed within the range of each predictor variable. To produce a maximum reduction in sum-of-squares residual error, the model adjusts the knot and its corresponding pair of BFs in each step. This process of adding BFs lasts and generally a very complex and overfitted model is produced. However, the overfitted model is pruned by deleting the less important redundant BFs in the backward stage [54, 55].

The MARS model $f(X)$ is generally expressed by the following equation;

$$f(x) = \delta_o + \sum_{m=1}^{M} \delta_m h_m(X) \tag{2}$$

Where δ_o and δ_m denote the coefficients which are calculated by the least sum of squared errors from splines functions, whereas $h_m(X)$ represents the spline functions, and M denotes the number of functions. The pruning stage improves the forecasting accuracy of the model and M is determined during this phase [55].

3.4 Model setup

The current study compares the accuracy of two machine learning methods such as MARS and M5Tree, for the prediction of daily relative humidity using different input data combinations of precipitation, temperature, and relative humidity. These machine learning models were applied on three meteorological stations such as Khunjerab, Naltar, and Ziarat one by one. The flowchart of the current study is displayed in **Figure 2**. Each model was applied on these stations separately with different input data combinations for the prediction of relative humidity (RH). Ten input data combinations were developed for each meteorological station by each model to decide the best input data combination for the prediction of relative humidity. Initially, three preceding relative humidity (RH) input combinations such as (i) RHt-1, (ii) RHt-1 and RHt-2, and (iii) RHt-1, RHt-2, and RHt-3 were tried to both the models to predict current RH (RHt). After that, three precipitation (i.e. (i) Pt-1, (ii) Pt-1, Pt-2, (iii) Pt-1,Pt-2,Pt-3) and temperature inputs (i.e. (i) Tt-1, (ii) Tt-1,Tt-2, (iii) Tt-1,Tt-2,Tt-3) combinations were separately added to the best RH combination whereas in the last input combination (10th); best temperature and precipitation inputs were added together with the best RH input combination to see the combine effect of both parameters on model's accuracy in predicting relative humidity.

The current analysis involves daily data of precipitation, temperature, and relative humidity from 1995 to 2009. About 75% of input data i.e. from 1995 to 2006 was used for training whereas 25% of input data i.e. from 2007 to 2009 was used for testing in both machine learning models for prediction of relative humidity.

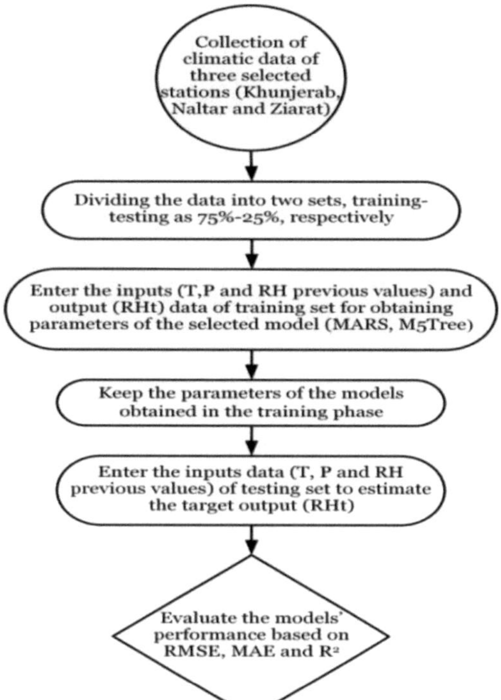

Figure 2.
Flowchart of the study.

However, [8] used only two-year data i.e. 2008 to 2009 for training the LSTM model which might not be enough for reliable predictions.

3.5 Models evaluation criteria

The models' accuracy in relative humidity prediction against observed data was evaluated using the following statistics which are normally used in the related literature. The statistics include R^2, RMSE, and MAE as shown in Eqs. (3)-(5).

$$R^2 = 1 - \frac{\frac{1}{n}\sum_{i=1}^{n}\left(RH_i - \overline{RH}\right)^2}{\frac{1}{n}\sum_{i=1}^{n}\left(rh_i - \overline{rh}\right)^2} \tag{3}$$

$$RMSE = \sqrt{\frac{\sum_{i=1}^{N}(RH_{iO} - RH_{iM})^2}{N}} \tag{4}$$

$$MAE = \frac{\sum_{i=1}^{N}|RH_{iO} - RH_{iM}|}{N} \tag{5}$$

Where $\overline{rh}$ indicates the observed mean relative humidity; $\overline{RH}$ is the mean of the predicted relative humidity RH_i; N signifies the number of data points. Moreover, RH_{iO} is observed relative humidity and RH_{iM} is modeled relative humidity. Previous studies such as [56–61] suggested that a single statistical indicator cannot examine well the prediction accuracy of soft computing models. Therefore, the current study used three statistical indicators to judge the model prediction accuracy with confidence. When the error distributions of the models are normal and uniform in that case the use of error statistics such as RMSE and MAE is more suitable. For an ideal model, the values of RMSE and MAE should equal to 0, whereas, R^2 should equal to 1. The model having relatively small values of MAE and RMSE as compared to other models is considered the best model.

4. Results and discussions

4.1 Performance evaluation of MARS model in predicting relative humidity

The performance evaluation statistics of the MARS model for the prediction of relative humidity at Khunjerab, Naltar, and Ziarat are presented in **Tables 3–5**, respectively. The MARS model performed excellent for the prediction of relative humidity at all meteorological stations both during training and testing processes especially, it provided the best predictions for the 6th scenario (S6) of input data combination which is highlighted in bold. The RMSE, MAE, and R^2 values during the training (5.58%, 4.51%, 0.852) and testing (5.98%, 5.43%, 0.808) stages for Khunjerab meteorological station are displayed in **Table 3**. The MARS model performed better during training as compared to testing at Khunjerab. However, the MARS model did not perform well for the S1, S2, and S3 scenarios. Our study results were found better than the study conducted by [1]. They described that GEP and ANNs models can predict relative humidity reliably at two Californian stations (RMSE= 10.7%, MAE= 7.6% and R^2 = 0.73) during training; and (RMSE= 10.1%, MAE= 7.5% and R^2 = 0.714) during testing stage in the case of GEP model. However, ANN model produced better results as compared to GEP such as (RMSE= 7.8%, MAE= 3.6% and R^2 = 0.826) during training, and (RMSE= 8.2%, MAE= 4.1% and R^2 = 0.751) during testing stage.

Scenario	Input Combinations	Training			Testing		
		RMSE (%)	MAE (%)	R^2	RMSE (%)	MAE (%)	R^2
S1	RHt-1	11.00	8.49	0.480	13.59	10.51	0.381
S2	RHt-1, RHt-2	10.88	8.39	0.491	13.58	10.51	0.385
S3	RHt-1, RHt-2, RHt-3	10.86	8.36	0.493	13.53	10.45	0.388
S4	RHt-1, RHt-2, RHt-3,Tt-1	5.78	4.64	0.823	6.43	5.62	0.782
S5	RHt-1, RHt-2, RHt-3,Tt-1,Tt-2	5.64	4.57	0.831	6.12	5.57	0.801
S6	**RHt-1, RHt-2, RHt-3,Tt-1,Tt-2,Tt-3**	**5.58**	**4.51**	**0.852**	**5.98**	**5.43**	**0.808**
S7	RHt-1, RHt-2, RHt-3,Pt-1	9.76	7.62	0.602	11.67	8.75	0.532
S8	RHt-1, RHt-2, RHt-3,Pt-1,Pt-2	9.16	7.28	0.643	10.87	8.38	0.544
S9	RHt-1, RHt-2, RHt-3,Pt-1,Pt-2,Pt-3	9.08	7.19	0.649	10.73	8.29	0.552
S10	RHt-1, RHt-2, RHt-3,Pt-1,Pt-2,Pt-3, Tt-1,Tt-2,Tt-3	6.21	5.13	0.802	6.03	5.53	0.803

Bold values represent the best input data combination.

Table 3.
The statistical evaluation of the MARS model at Khunjerab.

Scenario	Input Combinations	Training			Testing		
		RMSE (%)	MAE (%)	R^2	RMSE (%)	MAE (%)	R^2
S1	RHt-1	11.42	8.95	0.588	14.22	10.69	0.499
S2	RHt-1, RHt-2	11.09	8.65	0.612	13.99	10.31	0.518
S3	RHt-1, RHt-2, RHt-3	11.02	8.61	0.616	13.98	10.32	0.517
S4	RHt-1, RHt-2, RHt-3,Tt-1	5.84	4.73	0.812	6.84	5.34	0.783
S5	RHt-1, RHt-2, RHt-3,Tt-1,Tt-2	5.76	4.62	0.818	6.73	5.25	0.792
S6	**RHt-1, RHt-2, RHt-3,Tt-1,Tt-2,Tt-3**	**5.63**	**4.53**	**0.826**	**6.58**	**5.08**	**0.806**
S7	RHt-1, RHt-2, RHt-3,Pt-1	10.24	7.63	0.673	11.73	8.36	0.624
S8	RHt-1, RHt-2, RHt-3,Pt-1,Pt-2	10.08	7.46	0.692	11.29	8.07	0.645
S9	RHt-1, RHt-2, RHt-3,Pt-1,Pt-2,Pt-3	9.36	7.13	0.724	10.76	7.93	0.663
S10	RHt-1, RHt-2, RHt-3,Pt-1,Pt-2,Pt-3, Tt-1,Tt-2,Tt-3	5.71	4.71	0.815	6.74	5.18	0.796

Bold values represent the best input data combination.

Table 4.
The statistical evaluation of the MARS model at Naltar.

Similarly, the MARS model provided the best prediction of relative humidity for the S6 input data scenario at Naltar both during training and testing stages as shown in **Table 4**. The RMSE, MAE and R^2 values for the best input parameter combination were 5.63%, 4.53%, and 0.826 respectively, during training whereas 6.58%, 5.08%, and 0.806, were during testing (**Table 4**). The MARS model did not perform well for S1, S2, and S3 input combinations. However, a study conducted by [5] observed that the XGBoost model provided the best prediction of relative humidity (MAE= 2.29%) as compared to SARIMA (MAE= 2.97%) and HW additive (MAE= 2.74%).

Scenario	Input Combinations	Training			Testing		
		RMSE (%)	MAE (%)	R^2	RMSE (%)	MAE (%)	R^2
S1	RHt-1	11.29	8.58	0.530	14.75	10.90	0.420
S2	RHt-1, RHt-2	11.17	8.49	0.540	14.73	10.86	0.424
S3	RHt-1, RHt-2, RHt-3	11.12	8.46	0.544	14.77	10.89	0.421
S4	RHt-1, RHt-2, RHt-3,Tt-1	5.73	4.92	0.801	6.13	5.23	0.792
S5	RHt-1, RHt-2, RHt-3,Tt-1,Tt-2	5.58	4.76	0.813	6.02	5.06	0.807
S6	**RHt-1, RHt-2, RHt-3,Tt-1,Tt-2,Tt-3**	**5.26**	**4.59**	**0.833**	**5.86**	**4.97**	**0.815**
S7	RHt-1, RHt-2, RHt-3,Pt-1	10.03	7.13	0.624	12.75	8.34	0.542
S8	RHt-1, RHt-2, RHt-3,Pt-1,Pt-2	9.75	6.48	0.687	11.78	8.07	0.568
S9	RHt-1, RHt-2, RHt-3,Pt-1,Pt-2,Pt-3	9.48	6.29	0.698	11.38	7.84	0.597
S10	RHt-1, RHt-2, RHt-3,Pt-1,Pt-2,Pt-3, Tt-1,Tt-2,Tt-3	5.38	4.68	0.820	5.94	5.02	0.812

Bold values represent the best input data combination.

Table 5.
The statistical evaluation of the MARS model at Ziarat.

However, the MARS model performed the best (RMSE= 5.86, MAE= 4.97%, R^2 = 0.815) for prediction of relative humidity at Ziarat for the S6 input combination during the testing stage as shown in **Table 5**. The MARS model also performed fairly well during training stage (RMSE= 5.26%, MAE= 4.59%, R^2 = 0.833) for S6 input combination. The MARS model provided a poor prediction of relative humidity for S1, S2, and S3 input scenarios (**Table 5**). Overall, the MARS model performed fairly well at Khunjerab (R^2= 0.852) and showed slightly low performance at Naltar (R^2 =0.826) for the S6 input combination during the training stage (**Tables 3–5**).

The MARS model performance was also evaluated by drawing scatter plots. The scatter plots had been drawn between observed and predicted relative humidity from 2007 to 2009 on daily data as displayed in **Figure 3**. Scatter plots also displayed that the MARS model outperformed for prediction of relative humidity at

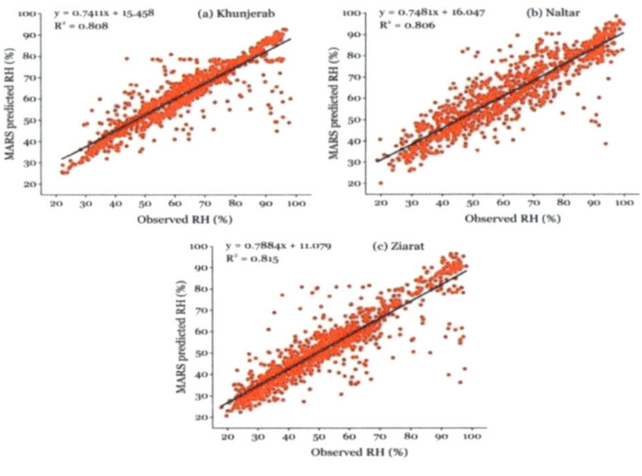

Figure 3.
Scatter plots between observed and predicted relative humidity by using MARS model at (a) Khunjerab; (b) Naltar and (c) Ziarat.

all meteorological stations, especially, at Ziarat with R^2 = 0.815 for the S6 input combination during the testing stage (**Figure 3**).

4.2 Performance evaluation of M5T model in predicting relative humidity

The performance evaluation of the M5T model for the prediction of relative humidity at Khunjerab, Naltar, and Ziarat is displayed in **Tables 6–8**, respectively. The M5T model also performed well for the prediction of relative humidity at all

Scenario	Input Combinations	Training			Testing		
		RMSE (%)	MAE (%)	R^2	RMSE (%)	MAE (%)	R^2
S1	RHt-1	11.08	8.56	0.476	13.64	10.54	0.378
S2	RHt-1, RHt-2	10.94	8.45	0.486	13.61	10.53	0.382
S3	RHt-1, RHt-2, RHt-3	10.90	8.41	0.491	13.56	10.49	0.391
S4	RHt-1, RHt-2, RHt-3,Tt-1	6.71	5.76	0.758	6.94	6.32	0.726
S5	RHt-1, RHt-2, RHt-3,Tt-1,Tt-2	6.62	5.64	0.765	6.89	6.13	0.748
S6	**RHt-1, RHt-2, RHt-3,Tt-1,Tt-2,Tt-3**	**5.94**	**5.08**	**0.796**	**6.14**	**5.56**	**0.772**
S7	RHt-1, RHt-2, RHt-3,Pt-1	9.79	7.68	0.598	11.77	8.82	0.529
S8	RHt-1, RHt-2, RHt-3,Pt-1,Pt-2	9.20	7.32	0.639	10.92	8.44	0.541
S9	RHt-1, RHt-2, RHt-3,Pt-1,Pt-2,Pt-3	9.13	7.24	0.642	10.83	8.36	0.548
S10	RHt-1, RHt-2, RHt-3,Pt-1,Pt-2,Pt-3, Tt-1,Tt-2,Tt-3	6.43	5.32	0.772	6.23	5.81	0.752

Bold values represent the best input data combination.

Table 6.
The statistical evaluation of the M5T model at Khunjerab.

Scenario	Input Combinations	Training			Testing		
		RMSE (%)	MAE (%)	R^2	RMSE (%)	MAE (%)	R^2
S1	RHt-1	11.12	8.57	0.476	13.68	10.58	0.378
S2	RHt-1, RHt-2	10.92	8.45	0.486	13.62	10.54	0.381
S3	RHt-1, RHt-2, RHt-3	10.89	8.38	0.491	13.57	10.48	0.384
S4	RHt-1, RHt-2, RHt-3,Tt-1	6.76	5.79	0.752	6.90	6.36	0.721
S5	RHt-1, RHt-2, RHt-3,Tt-1,Tt-2	6.58	5.61	0.760	6.81	6.18	0.742
S6	**RHt-1, RHt-2, RHt-3,Tt-1,Tt-2,Tt-3**	**5.82**	**5.12**	**0.791**	**6.19**	**5.58**	**0.762**
S7	RHt-1, RHt-2, RHt-3,Pt-1	9.84	7.67	0.598	11.76	8.79	0.529
S8	RHt-1, RHt-2, RHt-3,Pt-1,Pt-2	9.28	7.38	0.638	10.96	8.48	0.541
S9	RHt-1, RHt-2, RHt-3,Pt-1,Pt-2,Pt-3	9.11	7.23	0.646	10.77	8.32	0.550
S10	RHt-1, RHt-2, RHt-3,Pt-1,Pt-2,Pt-3, Tt-1,Tt-2,Tt-3	6.52	5.46	0.767	6.37	5.94	0.758

Bold values represent the best input data combination.

Table 7.
The statistical evaluation of the M5T model at Naltar.

Scenario	Input Combinations	Training			Testing		
		RMSE (%)	MAE (%)	R^2	RMSE (%)	MAE (%)	R^2
S1	RHt-1	11.12	8.54	0.476	13.64	10.56	0.378
S2	RHt-1, RHt-2	10.95	8.43	0.487	13.61	10.54	0.381
S3	RHt-1, RHt-2, RHt-3	10.92	8.39	0.491	13.58	10.49	0.384
S4	RHt-1, RHt-2, RHt-3,Tt-1	6.67	5.70	0.758	6.82	6.27	0.728
S5	RHt-1, RHt-2, RHt-3,Tt-1,Tt-2	6.47	5.52	0.764	6.72	6.10	0.752
S6	**RHt-1, RHt-2, RHt-3,Tt-1,Tt-2,Tt-3**	**5.74**	**5.04**	**0.796**	**6.08**	**5.46**	**0.783**
S7	RHt-1, RHt-2, RHt-3,Pt-1	9.79	7.66	0.599	11.72	8.78	0.530
S8	RHt-1, RHt-2, RHt-3,Pt-1,Pt-2	9.20	7.32	0.640	10.90	8.42	0.541
S9	RHt-1, RHt-2, RHt-3,Pt-1,Pt-2,Pt-3	9.12	7.21	0.645	10.78	8.32	0.550
S10	RHt-1, RHt-2, RHt-3,Pt-1,Pt-2,Pt-3, Tt-1,Tt-2,Tt-3	6.26	5.16	0.800	6.08	5.58	0.778

Bold values represent the best input data combination.

Table 8.
The statistical evaluation of the M5T model at Ziarat.

meteorological stations both during training and testing stages; however, it provided the best predictions of relative humidity for the 6th input data combination (S6) at all stations which are highlighted in bold. Overall, the M5T model performance was slightly lower as compared to MARS. The M5T model also performed better during training as compared to testing at all meteorological stations. However, the M5T model provided the best prediction of relative humidity at Ziarat as compared to Naltar and Khunjerab (**Table 8**). However, the M5T model did not perform well for the prediction of relative humidity for the S1, S2, and S3 scenarios with $R^2 < 0.50$ at all meteorological stations (**Tables 6–8**). A previous study conducted by [8] observed that the LSTM model is capable of forecasting complex univariate relative humidity time series. On contrary, [3] suggested that ARIMA can provide a better prediction of relative humidity as compared to LSTM.

At Khunjerab station, the M5T model performed well (RMSE= 5.94%, MAE = 5.08%, R^2= 0.796) in case of S6 input combination during model training stage whereas it displayed low prediction performance (RMSE= 6.14%, MAE= 5.56%, R^2= 0.772) during testing stage as shown in **Table 6**. Similarly, the M5T model did not perform well for the S1, S2, and S3 scenarios ($R^2 < 0.50$). Similarly, at Naltar station, the M5T model performed reasonably well (RMSE= 5.82%, MAE= 5.12%, R^2= 0.791) for S6 input combination during training stage whereas it exhibited a slightly low performance (RMSE= 6.19%, MAE= 5.58%, R^2= 0.762) during testing stage as presented in **Table 7**.

However, the M5T model provided the best prediction of relative humidity at the Ziarat station for the S6 input combination (**Table 8**). The M5T model performed better during training (RMSE= 5.74%, MAE= 5.04%, R^2= 0.796) as compared to testing (RMSE= 6.08%, MAE= 5.46%, R^2= 0.783) stage as displayed in **Table 8**.

The M5T model performance was also evaluated by drawing scatter plots. The scatter plots were drawn between observed and predicted relative humidity from 2007 to 2009 on daily data as displayed in **Figure 4**. Scatter plots showed that, the M5T model can also predict relative humidity fairly well at all meteorological stations, especially, at Ziarat (R^2= 0.782) for the S6 input combination during the testing stage (**Figure 4**).

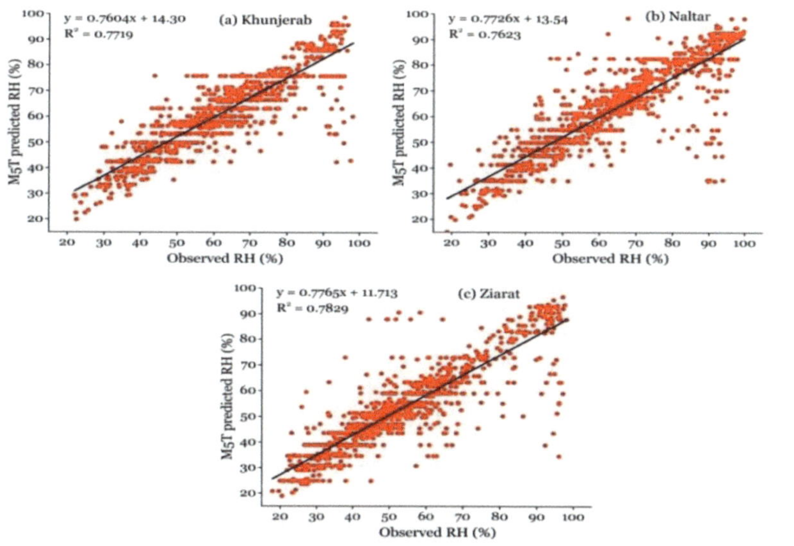

Figure 4.
Scatter plots between observed and predicted relative humidity by using the M5T model at (a) Khunjerab; (b) Naltar and (c) Ziarat.

4.3 Time variations of the observed and predicted relative humidity by MARS and M5T models

Time variations of the observed and predicted relative humidity by MARS and M5T model at Khunjerab, Naltar, and Ziarat meteorological stations are displayed in **Figures 5–7**. Time variations plots have been drawn by using the best-predicted data of relative humidity (i.e. S6 scenario). The daily data has been drawn from 2007 to 2009. **Figure 5** showed that both the models captured time-series variations of predicted relative humidity very well with reference to observed data at Khunjerab station but slightly underestimated the values from 900 to 1100 days. Moreover, these models slightly underestimated the low and high values of predicted relative humidity with reference to observed data at few points throughout the time series. Overall, the MARS model performed better as compared to M5T for the prediction of daily relative humidity data at Khunjerab.

The MARS and M5T models also captured time-series variation of relative humidity superbly with respect to observed data at Naltar station for the S6 input

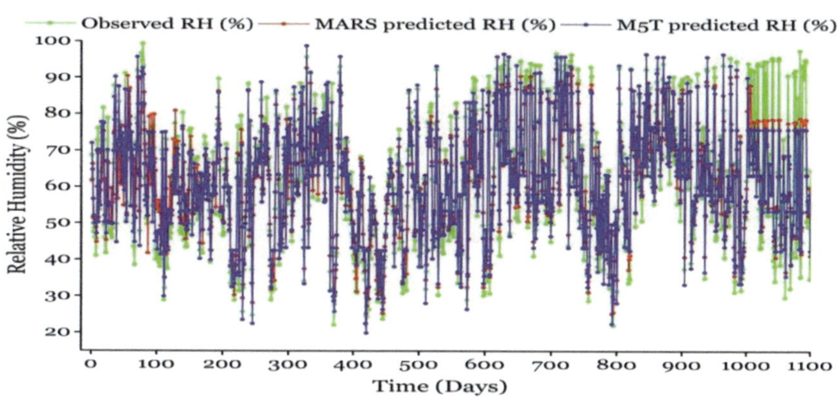

Figure 5.
Time variation of the observed and predicted relative humidity by MARS and M5T model at Khunjerab station.

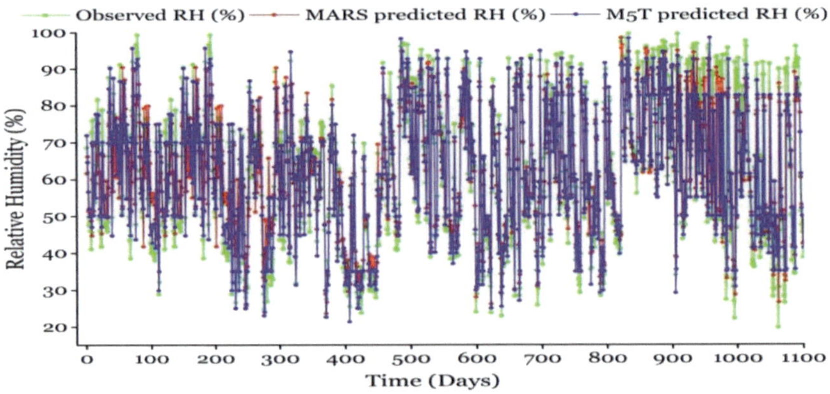

Figure 6.
Time variation of the observed and predicted relative humidity by MARS and M5T model at Naltar station.

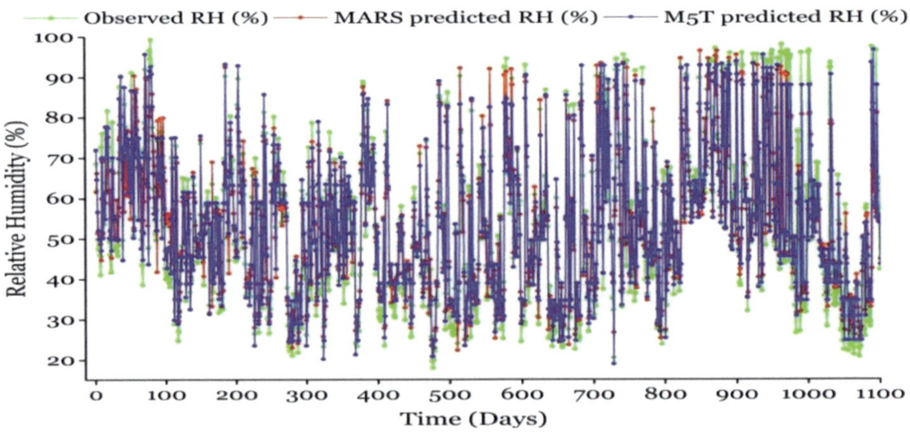

Figure 7.
Time variation of the observed and predicted relative humidity by MARS and M5T model at Ziarat station.

combination as displayed in **Figure 6**. Both the models slightly underestimated the predicted relative humidity from 850 days to 1100. Moreover, these models slightly underestimated the predictions of low and high values of relative humidity at some points throughout the study period. Overall, the MARS model provided better predictions of relative humidity as compared to M5T at Naltar (**Figure 6**).

However, both the machine learning models provided the best prediction of relative humidity at Ziarat which is a mid-altitude meteorological station as shown in **Figure 7**. Both the models captured the temporal variations of relative humidity very well throughout the period with reference to observed data for the S6 input combination. Furthermore, the models underestimated the low and high values of predicted relative humidity with reference to observed data. The MARS model predicted low and high values of relative humidity fairly well but it slightly underestimated the values at few points throughout the study period. Overall, the MARS model provided better predictions of relative humidity as compared to M5T at Ziarat (**Figure 7**).

5. Conclusions

Relative humidity has an important impact on plant growth, human health, industry, weather, and climate. Any change in temperature and relative humidity may result in droughts, heatwaves, floods, and hurricanes. Thus the relative humidity is one of the important factors to measure environmental changes. Keeping in view the importance of relative humidity, the current study has attempted to predict the relative humidity in a high elevated alpine basin (Hunza) of western Karakoram by using the MARS and M5T machine learning models. The current study is novel in that respect that previously nobody tried to predict the relative humidity in a high elevation alpine basin.

Statistical analysis of the model outputs suggested that both the models produced reliable predictions of relative humidity at Khunjerab, Naltar, and Ziarat meteorological stations of the Hunza basin during both training and testing stages. Out of 10 input data combinations of temperature, precipitation, and relative humidity, the 6th combination (i.e. RHt-1, RHt-2, RHt-3, Tt-1, Tt-2, Tt-3) produced the best results for each station by each model. The statistical indicators confirmed the excellent performance of both the models at all stations. For the MARS model, RMSE, MAE, and R^2 values ranged from 5.26–5.63%, 4.51–4.59%, and 0.826–0.856, respectively, during the training stage while they ranged from 5.86–6.58%, 4.97–5.43%, and 0.806–0.815, respectively, during the testing stage. However, in the case of the M5T model, the RMSE, MAE, and R^2 values ranged from 5.74–5.94%, 5.04–2.12%, and 0.791–0.796, respectively, during the training stage whereas the values ranged from 6.08–6.19%, 5.46–5.58%, and 0.762–0.783, respectively, during the testing stage of M5T model. Both the models showed poor performance such as (R^2 <0.50) in the case of S1, S2, and S3 input combinations at all stations. Moreover, it was observed that both the models performed better in training as compared to the testing stage. Both the models outperformed at Ziarat as compared to other stations. Overall, the MARS model performed better than M5T at all stations. The current study is important and it will provide a baseline for future studies to predict the other meteorological variables such as temperature, wind speed, solar radiation, and evapotranspiration by using machine learning tools in high altitude and remote basins which face the issue of data scarcity.

Acknowledgements

This study is supported by the Second Tibetan Plateau Scientific Expedition and Research Program (STEP, Grant No. 2019QZKK0208), and the Research Fund for Introducing Talents of Yunnan University (No. YJRC3201702). We are thankful to the Surface Water Hydrology Project of the Water and Power Development Authority (SWHP-WAPDA) of Pakistan for providing the required meteorological data to conduct this study.

Conflict of interest

The authors declare no conflict of interest.

Author details

Muhammad Adnan[1], Rana Muhammad Adnan[2], Shiyin Liu[1*], Muhammad Saifullah[3], Yasir Latif[4] and Mudassar Iqbal[5]

1 Institute of International Rivers and Eco-Security, Yunnan University, Kunming, China

2 State Key Laboratory of Hydrology-Water Resources and Hydraulic Engineering, College of Hydrology and Water Resources, Hohai University, Nanjing, China

3 Department of Agriculture Engineering, Muhammad Nawaz Sharif University of Agriculture, Multan, Pakistan

4 Key Laboratory of Tibetan Environment Changes and Land Surface Processes, Institute of Tibetan Plateau Research, Chinese Academy of Sciences, Beijing, China

5 Centre of Excellence in Water Resources Engineering, University of Engineering and Technology, Lahore, Pakistan

*Address all correspondence to: shiyin.liu@ynu.edu.cn

References

[1] Khatibi R, Naghipour L, Ghorbani MA, Aalami MT. Predictability of relative humidity by two artificial intelligence techniques using noisy data from two Californian gauging stations. Neural Computing and Applications. 2013; **23**(7):2241-52.

[2] Ghadiri M, Marjani A, Mohammadinia S, Shokri M. Machine Learning Approaches for Accurate Prediction of Relative Humidity based on Temperature and Wet-Bulb Depression. 2020.

[3] Li Z, Zou H, Qi B. Application of ARIMA and LSTM in Relative Humidity Prediction. In 2019 IEEE 19th International Conference on Communication Technology (ICCT). 2019; **16**:1544-1549.

[4] Azzouni A, Pujolle G. A long short-term memory recurrent neural network framework for network traffic matrix prediction. arXiv preprint. 2017; arXiv: 1705.05690.

[5] Li H, Yang Y, Cheng Y, Jin Y, Luo H, Zhang L. Application of Time Series Model in Relative Humidity Prediction. In Journal of Physics: Conference Series. 2020; **1584** (1):012017.

[6] Gangishetty MK, Scott RW, Kelly TL. Effect of relative humidity on crystal growth, device performance and hysteresis in planar heterojunction perovskite solar cells. Nanoscale. 2016; **8** (12):6300-7.

[7] Yu X. Indication of relative humidity of ECMWF in precipitation forecast in Hainan Prefecture. Qinghai Meteorology. 2009; **3**:17-20.

[8] Hutapea MI, Pratiwi YY, Sarkis IM, Jaya IK, Sinambela M. Prediction of relative humidity based on long short-term memory network. In AIP Conference Proceedings. 2020. **2221** (1): 060003.

[9] Quansah E, Amekudzi LK, Preko K. The influence of temperature and relative humidity on indoor ozone concentrations during the Harmattan. Journal of Emerging Trends in Engineering and Applied Sciences. 2012; **3**(5):863-7.

[10] Ohno H, Ohata T, Higuchi K. The influence of humidity on the ablation of continental-type glaciers. Annals of Glaciology. 1992; **16**:107-14.

[11] Hastenrath S. Recession of tropical glaciers. Science. 1994; **265**(5180): 1790-1.

[12] Kaser G, Hardy DR, Mölg T, Bradley RS, Hyera TM. Modern glacier retreat on Kilimanjaro as evidence of climate change: observations and facts. International Journal of Climatology: A Journal of the Royal Meteorological Society. 2004; **24**(3):329-39.

[13] Mölg T, Hardy DR. Ablation and associated energy balance of a horizontal glacier surface on Kilimanjaro. Journal of Geophysical Research: Atmospheres. 2004; **109**(D16).

[14] Rupper S, Roe G. Glacier changes and regional climate: A mass and energy balance approach. Journal of Climate. 2008; **21**(20):5384-401.

[15] Farhat N. Effect of relative humidity on evaporation rates in Nabatieh region. Lebanese Science Journal. 2018; **19**(1):59.

[16] Molano-Jimenez A, Orjuela-Cañón AD, Acosta-Burbano W. Temperature and Relative Humidity Prediction in Swine Livestock Buildings. In 2018 IEEE Latin American Conference on Computational Intelligence (LA-CCI). 2018; 1-4.

[17] Shi X, Lu W, Zhao Y, Qin P. Prediction of indoor temperature and relative humidity based on cloud

database by using an improved BP neural network in Chongqing. IEEE Access. 2018; **6**:30559-66.

[18] Singh VK, Tiwari KN. Prediction of greenhouse micro-climate using artificial neural network. Applied Ecology and Environmental Research. 2017; **15**(1):767-78.

[19] Gunawardhana LN, Al-Rawas GA, Kazama S. An alternative method for predicting relative humidity for climate change studies. Meteorological Applications. 2017; **24**(4):551-9.

[20] Bakar SZ, Ghazali RB, Ismail LH. Implementation of modified cuckoo search algorithm on functional link neural network for temperature and relative humidity prediction. In Proceedings of the First International Conference on Advanced Data and Information Engineering (DaEng-2013). 2014; 151-158.

[21] Martínez-Martínez V, Baladrón C, Gomez-Gil J, Ruiz-Ruiz G, Navas-Gracia LM, Aguiar JM, Carro B. Temperature and relative humidity estimation and prediction in the tobacco drying process using artificial neural networks. Sensors; **12**(10):14004-21.

[22] Özbalta TG, Sezer A, Yıldız Y. Models for prediction of daily mean indoor temperature and relative humidity: education building in Izmir, Turkey. Indoor and Built Environment. 2012; **21**(6):772-81.

[23] Chou SM, Lee TS, Shao YE, Chen IF. Mining the breast cancer pattern using artificial neural networks and multivariate adaptive regression splines. Expert Systems with Applications. 200; **27**(1):133-42.

[24] Adnan RM, Petroselli A, Heddam S, Santos CA, Kisi O. Short term rainfall-runoff modelling using several machine learning methods and a conceptual event-based model. Stochastic Environmental Research and Risk Assessment. 2021; **35**(3):597-616.

[25] Singh KK, Pal M, Singh VP. Estimation of mean annual flood in Indian catchments using back propagation neural network and M5 model tree. Water Resources Management. 2010; **24**(10):2007-19.

[26] Adnan RM, Liang Z, Trajkovic S, Zounemat-Kermani M, Li B, Kisi O. Daily streamflow prediction using optimally pruned extreme learning machine. Journal of Hydrology. 2019; **577**:123981.

[27] Sharda VN, Prasher SO, Patel RM, Ojasvi PR, Prakash C. Performance of Multivariate Adaptive Regression Splines (MARS) in predicting runoff in mid-Himalayan micro-watersheds with limited data/Performances de régressions par splines multiples et adaptives (MARS) pour la prévision d'écoulement au sein de micro-bassins versants Himalayens d'altitudes intermédiaires avec peu de données. Hydrological Sciences Journal. 2008; **53** (6):1165-75.

[28] Adamowski J, Chan HF, Prasher SO, Sharda VN. Comparison of multivariate adaptive regression splines with coupled wavelet transform artificial neural networks for runoff forecasting in Himalayan micro-watersheds with limited data. Journal of Hydroinformatics. 2012; **14**(3):731-44.

[29] Štravs L, Brilly M. Development of a low-flow forecasting model using the M5 machine learning method. Hydrological Sciences Journal. 2007; **52** (3):466-77.

[30] Sattari MT, Pal M, Apaydin H, Ozturk F. M5 model tree application in daily river flow forecasting in Sohu Stream, Turkey. Water Resources. 2013; **40**(3):233-42.

[31] Yaseen ZM, Kisi O, Demir V. Enhancing long-term streamflow

forecasting and predicting using periodicity data component: application of artificial intelligence. Water Resources Management. 2016(b); **30** (12):4125-51.

[32] Yin Z, Feng Q, Wen X, Deo RC, Yang L, Si J, He Z. Design and evaluation of SVR, MARS and M5Tree models for 1, 2 and 3-day lead time forecasting of river flow data in a semiarid mountainous catchment. Stochastic Environmental Research and Risk Assessment. 2018; **32**(9):2457-76.

[33] Nourani V, Davanlou Tajbakhsh A, Molajou A, Gokcekus H. Hybrid wavelet-M5 model tree for rainfall-runoff modeling. Journal of Hydrologic Engineering. 2019; **24**(5):04019012.

[34] Al-Sudani ZA, Salih SQ, Yaseen ZM. Development of multivariate adaptive regression spline integrated with differential evolution model for streamflow simulation. Journal of Hydrology. 2019; **573**:1-2.

[35] Mehdizadeh S, Fathian F, Safari MJ, Adamowski JF. Comparative assessment of time series and artificial intelligence models to estimate monthly streamflow: a local and external data analysis approach. Journal of Hydrology. 2019; **579**:124225.

[36] Adnan RM, Liang Z, Heddam S, Zounemat-Kermani M, Kisi O, Li B. Least square support vector machine and multivariate adaptive regression splines for streamflow prediction in mountainous basin using hydro-meteorological data as inputs. Journal of Hydrology. 2020; **586**:124371.

[37] Fathian F, Mehdizadeh S, Sales AK, Safari MJ. Hybrid models to improve the monthly river flow prediction: Integrating artificial intelligence and non-linear time series models. Journal of Hydrology. 2019; **575**:1200-13.

[38] Kisi O, Heddam S. Evaporation modelling by heuristic regression approaches using only temperature data. Hydrological Sciences Journal. 2019; **64**(6):653-72.

[39] Kaya YZ, Mamak M, Üneş F, Demirci M. Evapotranspiration prediction using M5T method and Ritchie equation for St. Johns, FL, USA. 2017.

[40] Alipour A, Yarahmadi J, Mahdavi M. Comparative study of M5 model tree and artificial neural network in estimating reference evapotranspiration using MODIS products. Journal of Climatology. 2014; **2014**.

[41] Shrestha S, Nepal S. Water balance assessment under different glacier coverage scenarios in the Hunza Basin. Water. 2019; **11**(6):1124.

[42] Saifullah M, Liu S, Adnan M, Ashraf M, Zaman M, Hashim S, Muhammad S. Risks of Glaciers Lakes Outburst Flood along China Pakistan Economic Corridor. In Glaciers and Polar Environment. 2020. IntechOpen.

[43] Hewitt K. Glacier change, concentration, and elevation effects in the Karakoram Himalaya, Upper Indus Basin. Mountain Research and Development. 2011; **31**(3):188-200.

[44] Ali AF, Zhang XP, Adnan M, Iqbal M, Khan G. Projection of future streamflow of the Hunza River Basin, Karakoram Range (Pakistan) using HBV hydrological model. Journal of Mountain Science. 2018; **15**(10):2218-35.

[45] Tahir AA, Chevallier P, Arnaud Y, Ahmad B. Snow cover dynamics and hydrological regime of the Hunza River basin, Karakoram Range, Northern Pakistan. Hydrology and Earth System Sciences. 2011; **15**(7):2275-90.

[46] Shrestha M, Koike T, Hirabayashi Y, Xue Y, Wang L, Rasul G, Ahmad B. Integrated simulation of snow and

glacier melt in water and energy balance based, distributed hydrological modeling framework at Hunza River Basin of Pakistan Karakoram region. Journal of Geophysical Research: Atmospheres. 2015; 120(10):4889-919.

[47] Quinlan JR. Learning with continuous classes. In 5th Australian joint conference on artificial intelligence. 1992; **92**:343-348.

[48] Witten IH, Frank E, Hall MA, Pal CJ. Practical machine learning tools and techniques. Morgan Kaufmann. 2005:578.

[49] Rahimikhoob A, Asadi M, Mashal M. A comparison between conventional and M5 model tree methods for converting pan evaporation to reference evapotranspiration for semi-arid region. Water Resources Management. 2013; **27**(14):4815-26.

[50] Adnan RM, Petroselli A, Heddam S, Santos CA, Kisi O. Comparison of different methodologies for rainfall–runoff modeling: machine learning vs. conceptual approach. Natural Hazards. 2021; **105**(3):2987-3011.

[51] Adnan RM, Liang Z, Yuan X, Kisi O, Akhlaq M, Li B. Comparison of LSSVR, M5RT, NF-GP, and NF-SC models for predictions of hourly wind speed and wind power based on cross-validation. Energies. 2019; **12**(2):329.

[52] Pal M, Deswal S. M5 model tree based modelling of reference evapotranspiration. Hydrological Processes: An International Journal. 2009; **23**(10):1437-43.

[53] Friedman JH. Multivariate adaptive regression splines. The Annals of Statistics. 1991:1-67.

[54] Zhang W, Goh AT. Multivariate adaptive regression splines and neural network models for prediction of pile drivability. Geoscience Frontiers. 2016; 7(1):45-52.

[55] Kisi O, Parmar KS. Application of least square support vector machine and multivariate adaptive regression spline models in long term prediction of river water pollution. Journal of Hydrology. 2016; **534**:104-12.

[56] Liu Z, Zhou P, Chen G, Guo L. Evaluating a coupled discrete wavelet transform and support vector regression for daily and monthly streamflow forecasting. Journal of Hydrology. 2014; **519**:2822-31.

[57] Yuan X, Wu X, Tian H, Yuan Y, Adnan RM. Parameter identification of nonlinear Muskingum model with backtracking search algorithm. Water Resources Management. 2016; **30**(8): 2767-83.

[58] Meng E, Huang S, Huang Q, Fang W, Wu L, Wang L. A robust method for non-stationary streamflow prediction based on improved EMD-SVM model. Journal of Hydrology. 2019; **568**:462-78.

[59] Zhou Y, Guo S, Chang FJ. Explore an evolutionary recurrent ANFIS for modelling multi-step-ahead flood forecasts. Journal of Hydrology. 2019; **570**:343-55.

[60] Alizamir M, Kisi O, Muhammad Adnan R, Kuriqi A. Modelling reference evapotranspiration by combining neuro-fuzzy and evolutionary strategies. Acta Geophysica. 2020; **68:**1113-26.

[61] Yuan X, Chen C, Lei X, Yuan Y, Adnan RM. Monthly runoff forecasting based on LSTM–ALO model. Stochastic environmental research and risk assessment. 2018; **32**:2199-212.

Section 2

Weather and Climate Change

Chapter 5

Time Trends and Persistence in the Snowpack Percentages by Watershed in Colorado

Luis Alberiko Gil-Alana

Abstract

In this paper we investigate the time trend coefficients in snowpack percentages by watershed in Colorado, US, allowing for the possibility of long range dependence or long memory processes. Nine series corresponding to the following watersheds are examined: Arkansas, Colorado, Gunnison, North Platte, Rio Grande, South Platte, San Juan-Animas-Dolores-San Miguel, Yampa & White and Colorado Statewide, based on annual data over the last eighty years. The longest series start in 1937 and all end in 2019. The results indicate that most of the series display a significant decline over time, showing negative time trend coefficients, and thus supporting the hypothesis of climate change and global warming. Nevertheless, there is no evidence of a long memory pattern in the data.

Keywords: snowpack percentages, time trends, long memory, Colorado, watersheds

1. Introduction

It is a well known fact that temperatures have been increasing during the last 50 years not only at global level but also at specific locations all over the world. In this paper we examine the statistical properties of nine time series corresponding to the snowpack percentages in watersheds in Colorado, US. Using annual data dating back to 1937, we are interested in the long memory feature of the data along with the time trend coefficients to check if the snowpack percentages have been declining in the last eighty years as a consequence of the effects of global warming. In addition, we test this hypothesis under the assumption that the underlying series display a long memory property, a feature that is very common in climatological data. As far as we know there are no previous works dealing with the statistical modeling of snow packs with time series data. Our results, using fractional integration, show no evidence of long memory, and the time trend coefficients of the snow packs are statistically significantly negative in the majority of the series examined, supporting thus the global warming hypothesis.

The standard approach to test for significant trends in time series is to consider a linear regression model of the following form:

$$y_t = \alpha + \beta t + x_t, \tag{1}$$

where a significant slope coefficient for β implies the presence of a trend (positive or negative, depending on the sign of the coefficient). However, in order to get consistent estimates of the unknown coefficients in (1), this set-up implicitly assumes that the error term, x_t must be well behaved and more specifically, it must be integrated of order 0 or I(0). This is a standard regularity condition and indicates that the infinite sum of its autocovariance values should be finite. This condition, however, is not always satisfied. If that sum is infinite, the series is said to be long memory, a feature widely observed in time series in many different disciplines including geophysical and climatological series, e.g., Beran [1], Percival et al. [2], Gil-Alana [3, 4], Ercan et al. [5], Graves et al. [6], etc.

In this article, this long memory feature is incorporated in our set-up by means of using a fractional integration model, which is described in the following section, and that is used to describe the error term x(t) in (1). Based on that, we test for the presence of significant time trends in the snowpack percentages at Colorado's watersheds.

The rest of the paper is structured as follows. Section 2 briefly describes the main idea of long memory or long range dependence processes and also presents the series under examination. Section 3 is devoted to the empirical results, while Section 4 concludes the manuscript.

2. Methodology and data: long memory

Given a covariance stationary process $\{x(t), t = 0, \pm 1, \dots\}$ we say that it is short memory or integrated of order 0 (and denoted as $x(t) \sim I(0)$) if the infinite sum of its autocovariances, defined as $\gamma(u) = Cov(x(t), x(t+u))$ is finite, i.e.,

$$\sum_{u=-\infty}^{\infty} \gamma(u) < \infty. \tag{2}$$

Within this category, we can include the white noise process but also the standard stationary AutoRegressive Moving Average (ARMA) type of models. This latter category allows for some type of dependence between the observations and is named "weak" (dependence) in the sense that the autocorrelations decay exponentially fast. However, many time series show higher degrees of dependence and belong to a category denominated as "long memory", characterized because the infinite sum of the autocovariances is infinite, i.e.,

$$\sum_{u=-\infty}^{\infty} \gamma(u) = \infty. \tag{3}$$

This long memory feature has been observed in time series data referring to many different disciplines, including economics and finance [7–9], energy [10–13], tourism [14, 15], environmental issues [16] and climatology [3, 17–19] among many others.

A very simple model, very popular among econometricians, and satisfying the above property (3) is the fractionally integrated or I(d, d > 0) model, which is expressed as:

$$(1 - L)^d x(t) = u(t), \quad t = 1, 2, \dots, \tag{4}$$

where L is the lag-operator, ie., $L^k x(t) = x(t-k)$, d can be any real positive value, and u(t) is I(0) or short memory as defined above. In this context, x(t) displays the property of long memory if $d > 0$. Using a Binomial expansion, the polynomial in the left-hand side in (4) can be expressed as:

$$(1-L)^d = 1 - dL + \frac{d(d-1)}{2}L^2 - \frac{d(d-1)(d-2)}{6}L^3 + \ldots \tag{5}$$

and x(t) in (4) can then be expressed as:

$$x(t) = dx(t-1) - \frac{d(d-1)}{2}x(t-2) + \frac{d(d-1)(d-2)}{6}x(t-3) + \ldots + \varepsilon(t) \tag{6}$$

and higher the value of d is, the higher the level of association between the observations is. A wide range of possibilities can be examined depending on the value of d in the real range. Examples are.

i. anti-persistence, if $d < 0$,

ii. short memory, if $d = 0$,

iii. stationary long memory processes, if $0 < d < 0.5$,

iv. nonstationary long memory mean reverting patterns, if $0.5 \leq d < 1$,

v. unit root processes, if $d = 1$, and

vi. explosive patterns with $d > 1$.

This specification is clearly more general than the standard methods used in the literature and that are based only on integer degrees of differentiation, i.e., d = 0 for stationarity and d = 1 for unit root or nonstationarity.

The series examined refer to the snowpack percentages in seven watersheds in Colorado, US, (see, **Figure 1**) namely, Arkansas, Colorado, Gunnison, North Platte, Rio Grande, San Juan-Animas-Dolores-and-San Miguel, South Platte, Yampa &

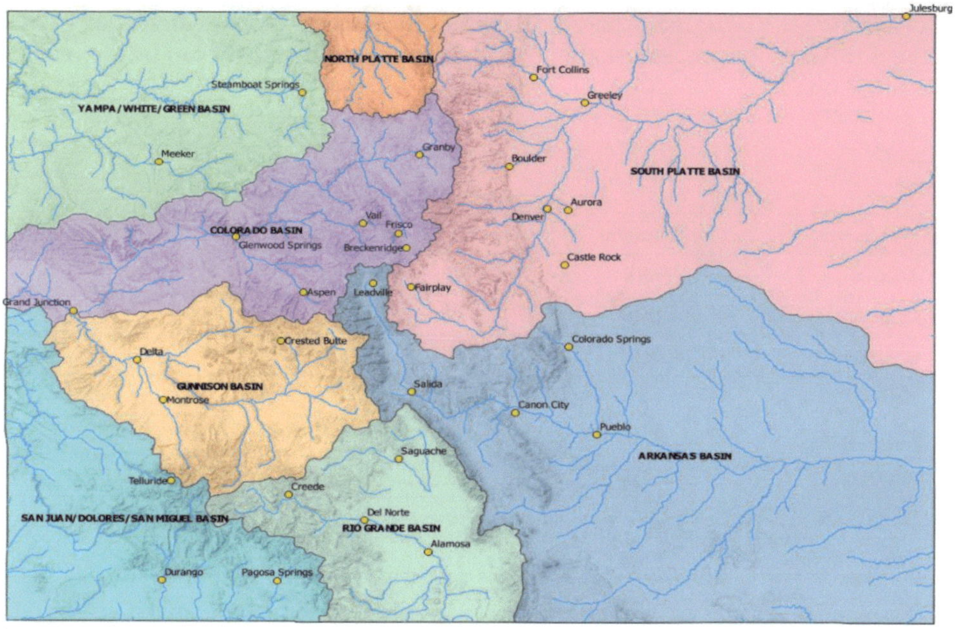

Figure 1.
Watersheds in Colorado, US. The different colors represent the different areas under study.

White and Colorado Statewide, using annual data. The starting year changes from one watershed to another and all of them end in 2019. The source of the data is the USDA (United States Department of Agriculture), Natural Resources Conservation Service Colorado (https://www.nrcs.usda.gov). All values use medians calculated from the most recent normal period between 1981 and 2010 For each basin we use data for the months of March, April and May, which are the months that present complete datasets for the whole period examined.

Table 1 displays the time series examined along with the acronym and the starting and ending years for each series. We observe that the long series are those referring to North Platte and Colorado Statewide (with data starting in 1937) and followed by South Platte and Colorado (starting in 1938), while the shortest one refers to San Juan, Animas, Dolores and San Miguel, starting in 1973.

The objective in this paper is twofold. First, to determine if long memory holds in the snowpack data examined, and second, to see if there is a decline in the time evolution of the data as a consequence of climate warming. Thus, in order to examine these two issues together, we consider the following model

$$y(t) = \alpha + \beta t + x(t); \quad (1 - L)^d x(t) = u(t), \tag{7}$$

where y(t) refers to the observed data, the snowpack percentages, α and β are unknown parameters referring respectively to a constant and a linear time trend, and the regressions errors x(t) are supposed to be I(d). Therefore, there are two main parameters in the above specification: β, related with the evolution over time of the series, and d, dealing with the short/memory feature of the data.

In the following section we display the estimated coefficients of d in Eq. (7) (along with their corresponding 95% confidence intervals) under three potential set-ups. The results presented in the second column in the table refer to the case where we impose α = β = 0 in (7); thus, no deterministic terms are included in the model. The results in column 3 refer to the model with an intercept, i.e. imposing β = 0 in (7). Finally, the last column reports the results where α and β are estimated from the data along with d.

We have marked the most appropriate case for each series in **Table 2** in bold. This selection has been made based on the significance of the estimated coefficients in the d-differenced process, noting that the two equations in (7) can be expressed in a single one as:

$$\tilde{y}(t) = \alpha \tilde{1}(t) + \beta \tilde{t}(t) + u(t), \tag{8}$$

Watershed	Acronym	Starting year	Ending year	n. of observ.
Arkansas	ARK	1950	2019	70
Colorado	COL	1938	2019	82
Gunnison	GUN	1941	2019	79
North Platte	NPB	1937	2019	83
Rio Grande	RIO	1950	2019	70
San Juan ...	SJ	1973	2019	47
South Platte	SPB	1938	2019	82
Yampa & White	YAM	1952	2019	68
Colorado statewide	YAM	1937	2019	83

San Juan ... refers to San Juan, Animas, Dolores and San Miguel.

Table 1.
Time series under examination.

where

$$\tilde{y}(t) = (1 - L)^d y(t), \tag{9}$$

$$\tilde{1}(t) = (1 - L)^d 1, \tag{10}$$

$$\tilde{t}(t) = (1 - L)^d t, \tag{11}$$

and noting that u(t) in (8) is I(0) by assumption, standard t-values hold here.

3. Results

We observe in **Table 2** that the time trend is required in 15 out of the 27 cases presented, while for the remaining 12, an intercept is sufficient to describe the deterministic part of the model. Of these 12 cases where only an intercept is required, five correspond to the month of March, another five occur in May and only two in April. Thus, the time trend is required in more than half of the series examined and as we will show below the coefficients are significantly negative in all cases implying that the snow packs are decreasing with time.

Watershed	Month	No terms	An intercept	A linear time trend
ARKANSAS BASIN	March	0.47 (0.20, 0.73)	0.02 (−0.13, 0.27)	**−0.16 (−0.40, 0.21)**
	April	0.47 (0.20, 0.74)	−0.02 (−0.15, 0.18)	**−0.28 (−0.47, 0.03)**
	May	−0.06 (−0.13, 0.48)	**−0.08 (−0.25, 0.18)**	−0.11 (−0.29, 0.17)
COLORADO BASIN	March	−0.05 (−0.08, 0.08)	**0.24 (−0.41, 0.01)**	−0.28 (−0.48, 0.00)
	April	0.08 (−0.08, 0.15)	−0.12 (−0.23, 0.06)	**−0.25 (−0.40, −0.02)**
	May	−0.07 (−0.11, 0.35)	**−0.19 (−0.36, 0.06)**	−0.19 (−0.37, 0.05)
GUNNISON BASIN	March	−0.06 (−0.11, 0.46)	**0.28 (−0.47, 0.00)**	−0.28 (−0.48, 0.00)
	April	−0.08 (−0.13, 0.41)	−0.11 (−0.23, 0.08)	**−0.20 (−0.35, 0.02)**
	May	−0.07 (−0.13, 0.29)	**−0.11 (−0.26, 0.13)**	−0.12 (−0.27, 0.12)
NORTH PLATTE BASIN	March	−0.03 (−0.06, 0.46)	**0.10 (−0.22, 0.09)**	−0.10 (−0.22, 0.09)
	April	0.09 (−0.08, 0.56)	**−0.05 (−0.17, 0.07)**	−0.11 (−0.21, 0.04)
	May	−0.06 (−0.10, 0.33)	**−0.09 (−0.20, 0.06)**	−0.13 (−0.25, 0.03)
RIO GRANDE BASIN	March	−0.13 (−0.21, 0.35)	−0.25 (−0.32, −0.06)	**−0.51 (−0.73, −0.22)**
	April	−0.08 (−0.12, 0.37)	**−0.12 (−0.24, 0.05)**	−0.17 (−0.29, 0.02)
	May	−0.02 (−0.10, 0.21)	**−0.03 (−0.15, 0.16)**	−0.04 (−0.16, 0.15)
SAN JUAN, ANIMAS, DOLORES AND SAN MIGUEL	March	−0.16 (−0.20, −0.07)	−0.27 (−0.44, 0.02)	**−0.53 (−0.82, −0.09)**
	April	0.00 (−0.23, 0.47)	0.00 (−0.13, 0.21)	**−0.44 (−0.73, −0.02)**
	May	−0.25 (−0.30, 0.51)	0.10 (−0.04, 0.31)	**−0.39 (−0.67, 0.07)**
SOUTH PLATTE BASIN	March	−0.06 (−0.10, 0.40)	**−0.09 (−0.22, 0.10)**	−0.15 (−0.31, 0.07)
	April	0.28 (0.15, 0.44)	0.04 (−0.06, 0.17)	**−0.16 (−0.31, 0.06)**
	May	0.27 (−0.12, 0.48)	0.03 (−0.09, 0.21)	**−0.16 (−0.35, 0.12)**
YAMPA & WHITE BASIN	March	0.23 (−0.04, 0.53)	0.18 (−0.31, 0.04)	**−0.37 (−0.63, −0.05)**
	April	0.04 (−0.21, 0.51)	−0.03 (−0.13, 0.12)	**−0.40 (−0.56, −0.14)**
	May	−0.06 (−0.33, 0.48)	−0.01 (−0.14, 0.18)	**−0.22 (−0.41, 0.08)**

Watershed	Month	No terms	An intercept	A linear time trend
COLORADO STATEWIDE BASIN	March	−0.04 (−0.09, 0.37)	**−0.19 (−0.33, 0.04)**	−0.21 (−0.37, 0.03)
	April	0.05 (−0.12, 0.45)	−0.04 (−0.14, 0.12)	**−0.18 (−0.32, 0.02)**
	May	−0.07 (−0.17, 0.36)	−0.06 (−0.19, 0.14)	**−0.15 (−0.30, 0.08)**

The values in parenthesis refer to the 95% confidence bands; in bold, the selected model for each series according to the deterministic terms.

Table 2.
Estimates of the differencing parameter.

Watershed	Month	No terms	An intercept	A linear time trend
ARKANSAS BASIN	March	−0.16 (−0.40, 0.21)	120.30 (27.37)	−0.338 (−2.96)
	April	−0.28 (−0.47, 0.03)	119.72 (43.53)	−0.348 (−4.65)
	May	−0.08 (−0.25, 0.18)	104.66 (29.77)	—
COLORADO BASIN	March	0.24 (−0.41, 0.01)	103.06 (102.35)	—
	April	−0.25 (−0.40, −0.02)	112.66 (48.55)	−0.153 (−2.86)
	May	−0.19 (−0.36, 0.06)	105.17 (57.61)	—
GUNNISON BASIN	March	0.28 (−0.47, 0.00)	106.76 (102.99)	—
	April	−0.20 (−0.35, 0.02)	114.57 (34.43)	−0.185 (−2.38)
	May	−0.11 (−0.26, 0.13)	108.27 (30.54)	—
NORTH PLATTE BASIN	March	0.10 (−0.22, 0.09)	104.10 (58.55)	—
	April	−0.05 (−0.17, 0.07)	106.25 (54.47)	—
	May	−0.09 (−0.20, 0.06)	106.45 (47.14)	—
RIO GRANDE BASIN	March	−0.51 (−0.73, −0.22)	112.80 (64.88)	−0.263 (−5.05)
	April	−0.12 (−0.24, 0.05)	99.769 (36.79)	—
	May	−0.03 (−0.15, 0.16)	92.624 (14.40)	—
SAN JUAN, ANIMAS, DOLORES AND SAN MIGUEL	March	−0.53 (−0.82, −0.09)	115.59 (44.98)	−0.461 (−4.03)
	April	−0.44 (−0.73, −0.02)	126.23 (38.08)	−1.005 (−7.06)
	May	−0.39 (−0.67, 0.07)	176.66 (26.53)	−2.556 (−9.12)
SOUTH PLATTE BASIN	March	−0.09 (−0.22, 0.10)	110.33 (49.03)	—
	April	−0.16 (−0.31, 0.06)	123.95 (34.76)	−0.337 (−4.26)
	May	−0.16 (−0.35, 0.12)	129.15 (26.99)	−0.391 (−3.68)
YAMPA & WHITE BASIN	March	−0.37 (−0.63, −0.05)	114.18 (57.22)	−0.225 (−3.88)
	April	−0.40 (−0.56, −0.14)	123.11 (71.75)	−0.441 (−8.73)
	May	−0.22 (−0.41, 0.08)	134.97 (22.75)	−0.663 (−4.09)
COLORADO STATEWIDE BASIN	March	−0.19 (−0.33, 0.04)	105.51 (83.54)	—
	April	−0.18 (−0.32, 0.02)	115.49 (38.53)	−0.222 (−3.35)
	May	−0.15 (−0.30, 0.08)	119.25 (22.40)	−0.267 (−2.30)

The values in parenthesis in the last two colums are their corresponding t-values.

Table 3.
Estimated coefficients of the selected models.

Table 3 displays the estimated coefficients of the selected models in **Table 2**. If we first focus on the values of d, we observe that there is no evidence of long memory behavior in any single case, since the confidence intervals include the value 0 in all cases examined. There are 21 series where the I(0) hypothesis of short memory cannot be rejected and in the remaining 6, anti-persistence ($d < 0$) is detected. The series showing anti-persistence are Colorado (April), Rio Grange (March) and San Juan ... and Yampa & White (in March and April). Anti-persistence is not as frequent as persistence, though some examples can be found in the literature [20, 21]. Anti-persistent behavior exhibits prolonged damped oscillations with the spectral density function showing a zero value at the origin [22].

If we look now at the time trend coefficients the first noticeable feature is that all the significant coefficients are negative, supporting the hypothesis of a decline in the snowpack percentage in a number of cases. The highest coefficients refer to the cases of San Juan, Animas, Dolores and San Miguel in May (with an estimated time trend coefficient of −2.556) and in April (−1.005). Other high significant negative coefficients are those of Yampa & White (in May, −0.663, and April, −0.441) and San Juan ... in March (−0.461). All these cases support the hypothesis of a decreasing trend in the snow packs in various Colorado's watersheds, which might be a consequence of the global warming climate hypothesis.

As a robustness method, we also employ alternative parametric and semiparametric methods of estimating the differencing parameter in the context of fractional integration, including among other Sowell's [23] maximum likelihood estimation method, the classical semiparametric Geweke and Porter-Hudak's [24] approach and the most recent developed method in Shimotsu [25] and the results support our conclusions in all cases, finding evidence of short memory and negative time trend coefficients in most of the series examined.

4. Conclusions

We have examined nine time series in this paper corresponding to snowpack percentages in Colorado, investigating if there has been a significant decline over time in the series in the context of long memory processes. For this purpose, we have tested for the significance of the time trend coefficient in a model where the regression errors are fractionally integrated or integrated of order d. Long memory occurs then if d is a positive value.

Our results indicate that there is no evidence of long memory behavior since all the orders of integration are close to zero or below it, implying short memory or anti-persistence behavior. Focusing on the time trend coefficients, these are significant in 15 of the 27 series examined, and in all these cases, they are found to be significantly negative, supporting thus the hypothesis of a decline in the amount of snow in Colorado watersheds probably as a consequence of global climate warming.

Future work should focus on alternative modeling approaches including for example non-linear structures which are clearly related with long memory and fractional integration models [26]. Thus, non-linear deterministic terms like those based on Chebyshev polynomials in time [27] or Fourier transforms [28], in both cases based on I(d) models, can also be implemented on these or on similar data.

Acknowledgements

The author gratefully acknowledges financial support from the MINEIC-AEI-FEDER ECO2017-85503-R project from 'Ministerio de Economía, Industria y

Competitividad' (MINEIC), 'Agencia Estatal de Investigación' (AEI) Spain and 'Fondo Europeo de Desarrollo Regional' (FEDER).

Comments from the Editor and an anonymous reviewer are also gratefully acknowledged.

Conflict of interest

The authors declare no conflict of interest.

Author details

Luis Alberiko Gil-Alana
University of Navarra, Pamplona, Spain

*Address all correspondence to: alana@unav.es

References

[1] Beran, J. (1994). Statistics for long-memory processes. Chapman & Hall: New York, NY, USA, 1–315.

[2] Percival, D.B., Overland, J.E. and Mofjeld, H.O. (2001) Interpretation of North Pacific variability as a short- and long-memory process. Journal of Climate, 14(24), 4545–4559.

[3] Gil-Alana, L.A, (2005), Statistical model of the temperatures in the northerm hemisphere using fractionally integrated techniques. Journal of Climate 18 (24), 5357–5369.

[4] Gil-Alana, L.A. (2006) Nonstationary, long memory and antipersistence in several climatological time series data. Environmental Modeling and Assessment, 11(1), 19–29.

[5] Ercan, A., Kavvas, M.L., Abbasov, R. K. (2013). Long-range dependence and sea level forecasting. Springer International Publishing, 1–51.

[6] Graves, T., Gramacy, R., Watkins, N., and Franzke, C. (2017). A brief history of long memory: Hurst, Mandelbrot and the road to ARFIMA, 1951–1980. Entropy, 19(9) 437, 1–21.

[7] Abbritti, M., L.A. Gil-Alana, Y. Lovcha and A. Moreno (2016). Term structure persistence, Journal of Financial Econometrics, 14, 2, 331–352.

[8] Gil-Alana, L.A. and A. Moreno (2012). Uncovering the U.S. term premium.An alternative route. Journal of Banking and Finance, 36, 1184–1193.

[9] Haug, A.. (2014), On Real Interest Rate Persistence: The Role of Breaks, Applied Economics, 46, 1058–1066

[10] Barros, C.P., L.A. Gil-Alana and P. Wanke, 2016, Energy production in Brazil. Empirical facts based on persistence, seasonality and breaks, Energy Economics 54, 88–95,

[11] Elder, J., Serletis, A.(2008). "Long memory in energy futures prices". Review of Financial Economics . *17* , 146–155

[12] Gil-Alana, L.A., H. Carcel and R. Mudida (2017), Shocks affecting electricity prices in Kenya. A fractional integration study, Energy 124, 1, 521–530.

[13] Solarin, S.A., L.A. Gil-Alana and C. Lafuente, 2020, An investigation of long range reliance on shale oil and shale gas production in the US market, Energy 195.

[14] Al-Shboul, M. and S. Anwar, 2017, Long memory behavior in Singapore's tourism market, International Journal of Tourism Research 19, 5, 524–534.

[15] Gil-Alana, L.A. and P. Huijbens (2018), Tourism in Iceland. Persistence, seasonality and long memory behaviour, Annals of Tourism Research 68, 20–29.

[16] Gil-Alana, L.A., S.A. Solarin and C. Lafuente (2019), Persistence in carbon footprint emissions. An overview over of 92 countries, Carbon Management 10, 4, 405–415.

[17] Gil-Alana, L.A. and M. Monge, (2020), Global CO_2 emissions and global temperatures. Are they related, International Journal of Climatology, forthcoming.

[18] Rea, W., M. Reale and J. Brown, 2011, Long memory in temperature reconstruction, Climatic Change 107, 247–265.

[19] Rodrigo-Caballero, S., J. Brix and A. Brix, 2020, Long memory in surface air temperatura, Climate Research 21, 127–140.

[20] Caporale, G. M., Gil-Alana, L. A., and Poza, C., (2020). Persistence, nonlinearities and structural breaks in European stock market indices. The Quarterly Review of Economics and Finance 77, 50–61.

[21] Dimitrova, V., Fernández-Martínez M., Sánchez-Granero, M. A., and Trinidad Segovia, J. E., (2019). Some comments on Bitcoin market (in) efficiency. Plos One 14, e0219243.

[22] Dittmann, I., and Granger, C. W. J., (2002). Properties of nonlinear transformations of fractionally integrated processes. Journal of Econometrics. 110, 113–133.

[23] Sowell, F., (1992), Maximum likelihood estimation of stationary univariate fractionally integrated time series models, Journal of Econometrics 53, 1–3, 165–188.

[24] Geweke, J. and S. Porter-Hudak, (1982), The estimation and application of long memory time series models, Journal of Time Series Analysis 4, 4, 221–238.

[25] Shimotsu, K. (2010): Exact Local Whittle Estimation Of Fractional Integration with Unknown Mean and Time Trend. Econometric Theory, Vol. 26, pp. 501–540.

[26] Diebold, F.X. and A. Inoue (2001), Long memory and regime switching, Journal of Econometrics 105, 131–159.

[27] Cuestas J.C. and L.A. Gil-Alana (2016). "Testing for long memory in the presence of non-linear deterministic tre nds with Chebyshev polynomials," Stud ies in Nonlinear Dynamics and Econome trics, De Gruyter, vol. 20(1), pages 57–74, February.

[28] Gil-Alana, L.A. and O. Yaya, 2020, Testing fractional unit roots with non-linear smooth break approximations using Fourier functions, Journal of Applied Statistics, DOI: 10.1080/02664763.2020.1757047

Chapter 6

Future Climate Change Impacts on River Discharge Seasonality for Selected West African River Basins

Toju Esther Babalola, Philip Gbenro Oguntunde, Ayodele Ebenezer Ajayi and Francis Omowonuola Akinluyi

Abstract

The changing climate is a concern to sustainable water resources. This study examined climate change impacts on river discharge seasonality in two West African river basins; the Niger river basin and the Hadejia-Jama'are Komadugu-Yobe Basin (HJKYB). The basins have their gauges located within Nigeria and cover the major climatic settings. Here, we set up and validated the hyper resolution global hydrological model PCR-GLOBWB for these rivers. Time series plots as well five performance evaluation metrics such as Kling–Gupta efficiency (KGE),); the ratio of RMSE-observations standard deviation (RSR); per cent bias (PBIAS); the Nash–Sutcliffe Efficiency criteria (NSE); and, the coefficient of determination (r2), were employed to verify the PCR-GLOBWB simulation capability. The validation results showed from satisfactory to very good on individual rivers as specified by PBIAS (−25 to 0.8), NSE (from 0.6 to 0.8), RSR (from 0.62 to 0.4), r2 (from 0.62 to 0.88), and KGE (from 0.69 to 0.88) respectively. The impact assessment was performed by driving the model with climate projections from five global climate models for the representative concentration pathways (RCPs) 4.5 and 8.5. We examined the median and range of expected changes in seasonal discharge in the far future (2070–2099). Our results show that the impacts of climate change cause a reduction in discharge volume at the beginning of the high flow period and an increase in discharge towards the ending of the high flow period relative to the historical period across the selected rivers. In the Niger river basin, at the Lokoja gauge, projected decreases added up to 512 m3/s under RCP 4.5 (June to July) and 3652 m3/s under RCP 8.5 (June to August). The three chosen gauges at the HJKYB also showed similar impacts. At the Gashua gauge, discharge volume increased by 371 m3/s (RCP8.5) and 191 m3/s (RCP4.5) from August to November. At the Bunga gauge, a reduction/increase of -91 m3/s/+84 m3/s (RCP 8.5) and -40 m3/s/+31 m3/s/(RCP 4.5) from June to July/August to October was simulated. While at the Wudil gauge, a reduction/increase in discharge volumes of −39/+133 m3/s (RCP8.5) and −40/133 m3/s (RCP 4.5) from June to August/September to December is projected. This decrease is explained by a delayed start of the rainy season. In all four rivers, projected river discharge seasonality is amplified under the high-end emission scenario (RCP8.5). This finding supports the potential advantages of reduced greenhouse gas emissions for the seasonal river discharge regime. Our study is anticipated to provide useful information to policymakers and river basin development authorities, leading to improved water management schemes within the context of changing climate and increasing need for agricultural expansion.

Keywords: west African rivers, seasonality, climate change, PCR-GLOBWB, river discharge

1. Introduction

Many regions worldwide have experienced modifications in their hydrological regimes due to changing global climate [1]. The changing climate significantly impacts the availability and quantity of water in individual river basins and, on a global scale, has long been an international concern [2]. Freshwater availability in West Africa is paramount to economic development and social well-being [3]. In most West African countries, agriculture is the crucial sector supporting about 60% of the region's population [4]. Agriculture production in Nigeria and other West African countries depends majorly on water availability. Most importantly, wetlands watersheds contribute in diverse ways to the livelihoods of millions of people in this region [5, 6]. However, due to climate change, West Africa is already experiencing sea-level rise with severe coastline erosion, increased temperatures, unpredictable rainfall, dwindling water resources and many more. Accordingly, it is critical to examine future consequences of climate change on river discharge seasonality for a more explicit and quantitative understanding of the available water for various functions. Knowledge about river discharge seasonality is necessary for understanding important interannual hydrological dynamics [7].

There exist intricate interactions between streamflow and other climate variables (e.g. precipitation, temperature, and evapotranspiration) in a catchment [8, 9]. Therefore, the effects of climate change on streamflow across a given region are often assessed using hydrological models forced with global climate models [10]. Many studies that have examined climate change impacts on discharge seasonality globally can be found in the literature. Dettinger and Diaz [11]; explored on a global scale, the seasonality and variability of streamflow. They concluded that local seasonal cycles of necessary climatic inputs influence streamflow seasonality. On a regional scale, Aich et al. [12] compared the impacts of climate change on streamflow regimes in four large African river basins and found visible impacts of climate change on high and low flows and mean discharges. Eisner et al. [13] investigated climate change impacts on streamflow seasonality, over eleven large river basins, using an ensemble of hydrological and climate models. They found an increasing/decreasing tendency for high and low flows in many of the basins. Hirpa et al. [14] estimated the response of future streamflow in the greater Horn of Africa, using a distributed hydrological model driven by an ensemble of climate models. The study found a reduction in streamflow in the major rivers in Ethiopia and increasing streamflow projections in the equatorial zone towards ending of the century. WaleWorqlul et al. [15] found that climate change would significantly influence the hydrology of two subbasins in the upper Nile in Ethiopia. Li and Jin [16] in their study also found increasing streamflow variability attributed to increasing rainfall variability in the Jing River of China. Ficklin et al. [17], in their study, projected an arid-conditions by the 2080s for many subbasins in the Upper Colorado River Basin of the southwestern United States when they investigated the impacts of Climate Change on Streamflow and hydrology of the basin. Vano et al. [18] identified seasons susceptible to future climate change in the Pacific Northwest using the Variable Infiltration Capacity hydrology mode forced with an ensemble of ten global climate models. These studies and many others all show that climate change plays an increasingly important role in river discharge seasonal flow; hence, it is crucial to know how the climate changes would affect seasonal variations of a basin's streamflow.

The vulnerability of most West African countries to climate change originates from high reliance on climate-based economic activities and large populations in coastal urban areas [19]. Climate change may undoubtedly modify the river flow regime of most West African basins. It may cause a temporary increase of flow in some places or reduce river flows in other areas [20]. Few studies have addressed climate change impacts on the seasonal river discharge regime in West Africa, especially over Nigeria. Ayeni et al. [20] assessed the impacts of climate changes on three basins of southwestern Nigeria for the period (1961–2007), using the Pitman rainfall-runoff model forced with CSIRO Mark3.5, MIROC3.2-medres and UKMO-HadCM3 GCMs. Ndulue and Mbajiorgu [21] quantified the hydrological changes due to climate impacts and land use in the Upper Ebonyi river watershed, south-eastern Nigeria, for the present period (1985–2014) and the future period (2020–2030 and 2040–2050), using the SWAT forced with the CSIRO-Mk3–6-0 climate model.

The availability of observation data has limited research in this region; the majority of the river basins are ungauged [21]. Also, studies in west African tend to be fragmentary [22], as they use different; climate models; bias correction approach, emission pathways; meteorological forcings; period considered (future and historical). Furthermore, as water availability is critical to economic and social well-being, it is vital to understand the relationship between climate change and streamflow and how it will affect streamflow to establish appropriate adaptation policies. These reasons, coupled with the region's high vulnerability, have made it crucial to provide reliable future projections for river discharge seasonality.

This study, therefore, aims to assess the climate change impacts on far-future river discharge seasonality. We used the PCR-GLOBWB model, forced with five consistent climate models, considering two future emission scenarios (RCP4.5 and RCP 8.5). This study is anticipated to provide useful information to policymakers and river basin development authorities, leading to the improvement of water management within the context of climate change. Our study will also help concerned river basin authorities to improve water management and future socio-economic development for river basins in Nigeria.

2. Study area

Two river basins, with at least five consecutive hydrological years record, over the period 1958–2015 were selected. River selection are from Global Runoff Data Center (GRDC)'s [23] available collection of West African basins. Gauges of the rivers are located in North-central and North-east (**Figure 1**) of Nigeria. Besides, the location of the gauges covers the vast tropical savannah climatic region, which covers most of the country (west to centre Nigeria) and the Arid climatic region found in the North [24]. Climate change impacts in Nigeria has been identified as disastrous due to the country's vulnerability and deficient coping capability [25]. The consequences are evidenced in rising temperature, variable precipitation, sea-level rise and flooding, desertification and drought, altered water resources, and biodiversity loss [26]. Moreover, the ecological system, agricultural systems, and livelihoods that these basins support are susceptible to the reality of climate change. The selected basins are briefly described in **Table 1**.

The Niger river basin: is the longest catchment in West Africa. The Niger watershed stretches over an expanse of terrain that spans ten west African countries, covering an area of around 2,156,000 km^2, with nearly 1,270,000 km^2 hydrologically active [27]. The Niger river comprises of different hydrographic regions uniquely identified by drainage and hydrological characteristics. The Guinea highland is the river source, flowing to the Sahara into Mali through the Inner Delta. From

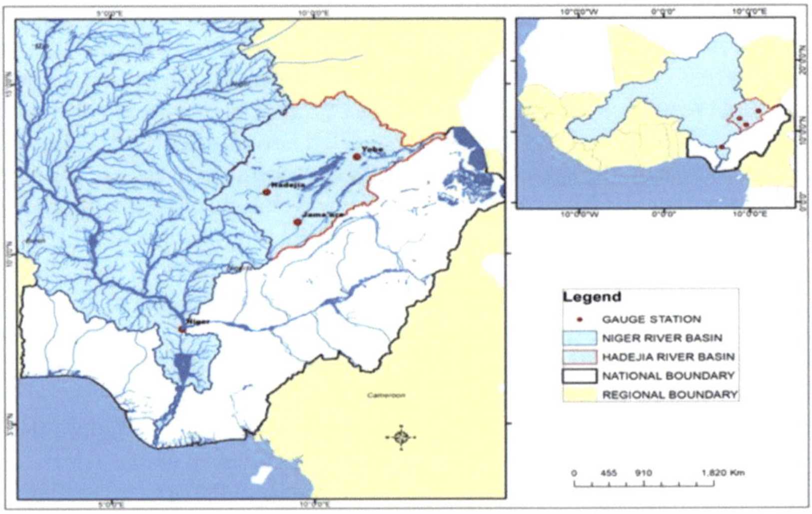

Figure 1.
Study area map.

River Basin	Niger	Hadejia-Jama'are Komadugu-Yobe Basin		
River name	Niger	Hadejia	Jama'are	Yobe
Gauge station	Lokoja	Wudil	Bunga	Gashua
Area (km^2)	2,156,000	16,380	7977	62,150
GRDC number	1,834,101	1,837,410	1,837,255	1,837,107
Annual precipitation (1971–2000) (mm)	1403	950	909	414
Mean temperature (1971–2000) (°C)	26.7	25.9	25.7	27.9
Koppen-Geiger climate classification	Tropical Savannah (Aw)	Tropical Savannah (Aw)	Tropical Savannah (Aw)	Arid Steppe hot (Bsh)

Table 1.
Description of selected rivers and gauges.

the Inner Delta, it flows southeast, and join with its major tributary, River Benue and eventually enters the Atlantic Ocean by the Niger Delta within Nigeria [28]. 28.3% (424,500 km2), of the hydrologically active basin falls within Nigeria (the lower Niger) and stretches over 20 of the country's 36 states. Within Nigeria, the Niger river constitutes two major rivers (the Niger and Benue) and more than half of its major rivers. Approximately 60% of the Nigerian population lives in the basin area. 71% of over 100 million people inhabiting the Niger basin live in Nigeria [28]. Agriculture, particularly rain-fed agriculture, supports most of the inhabitants' livelihood and is therefore susceptible to impacts of climate variability. The gauging station on the Niger river in this study is Lokoja at the confluence of the river Niger and Benue.

The Hadejia-Jama'are Komadugu-Yobe Basin (HJKYB) is a combined basin of the Komadugu-Yobe river and the Hadejia-Jama'are River System located in the north-east part of Nigeria. The Hadejia and Jama'are river are the two main rivers of this basin that meet at the Hadejia Nguru Wetlands (HNW) to form the Komadugu-Yobe Basin [29]. The Hadejia river has its source in the Kano highlands, while the Jama'are river rises from the Jos plateau in Nigeria. The combined catchment of the

HJKYB discharges into Lake Chad, and it is the major recharge into the lake from Nigeria. The HJKYB is one of the essential basins in Nigeria that support over 15 million people, which mostly rely on agriculture, fishing, livestock rearing, and water supply [30] for their livelihoods. An essential part of the HJKYB is the HNW which was the pride and joy of the north-eastern part of Nigeria for many years. The research site on the HJKYB is at the Gashua gauge station on the Komadugu-Yobe River, Bunga gauging station on the Jama'are River, and the Wudil on the Hadejia river.

2.1 Methods

2.1.1 The PCRaster global water balance model

The PCR-GLOBWB is a large scale hydrological model, which incorporates human activities into hydrology [31]. Global simulation of hydrology is daily and at a spatial resolution of 5 arcmin (0.08° at the equator). In this study, our focus is on river basins in West Africa, specifically for gauging stations located within Nigeria. The PCR-GLOBWB simulates water stored in two top soil layers S_1 and S_2; one bottom groundwater reservoir S_3 for each grid cell and time step. The model also simulates; water movement between the atmosphere and layer of topsoil (precipitation, evaporation, transpiration, and snowmelt); among the soil; in between the soil and the active layer of groundwater and estimates interception by canopy and snow storage. Distinct land cover types (forest, grassland, irrigated paddy field, irrigated non-paddy field, and open water), soil types, and elevation are considered to determine sub-grid variability. The Improved ARNO scheme is used in the model to estimate the fraction of the area of saturated soil [32]. Precipitation can be intercepted, evaporated, or infiltrated into the soil layers. The excess surface runoff (Q_{dr}), second soil layer (interflow) runoff (Q_{sf}), and groundwater (baseflow) runoff (Q_{bf}) make up the runoff from each cell. To obtain the river discharge (Qchannel), Specific runoff from each cell is gathered and then routed through the drainage network following the travel-time solution of [33]. At each time step, the model simulates (i) livestock, household irrigation and industrial water demand, (ii) water withdrawn from surface water, groundwater and desalinisation. For this work, we adopted the guideline of standard parameterisation of [31] of PCR-GLOBWB, which is based on available global datasets.

2.2 Data

2.2.1 Forcing datasets

Datasets, as described in [34], is followed and repeated in this study. The meteorological datasets required to drive the model are precipitation, temperature, and reference potential evapotranspiration. We obtained these data from the CRU TS 3.2 [34]. These data were processed by interpolating station observation past time-series to a global grid resolution of 0.5^0. Due to the daily resolution of PCR-GLOBWB, the monthly CRU TS 3.2 data were downscaled to daily resolution with ERA 40 (1958–1978, [35]) and ERA-Interim (1979–2015, [36]). ERA 40 and ERA-I had been spatially downscaled from their initial spatial resolutions of 1.2° and 0.7° to 0.5° in the resampling scheme of the European Centre for Medium-Range Weather Forecasts (ECMWF). This downscaling was done by first allotting the larger values ERA40 and ERA-I to the middle of the cells and then interpolating spatially to the higher resolution of 0.5°. Firstly, downscaling of precipitation was done by temporarily assigning a threshold of 0.1 mm day^{-1} to the daily time series of ERA, thereby estimating the number of days with rain and eliminating the drizzle effect. The rainfall quantity below this threshold was proportionally allocated to the rainy days. Thereafter, CRU monthly precipitation was reproduced by multiplicative scaling of the daily rainfall totals. Also, monthly reference potential evaporation, estimated from the CRU dataset with Penman-Monteith, was scaled using multiplicative scaling and downscaled to daily data using a daily temperature-based ET product derived from daily ERA temperatures. An additive scaling method was used for air temperature (see [31] for more details). For this work, standard parameterisation guideline, as provided in [31], was adopted. We used available global datasets, including vegetation, geological information, and soil properties, to parameterise the model and simulate discharge at daily time steps over the selected river basins (from 1958 to 2015). Monthly averages were used to report the output from the Model.

2.2.2 Discharge data

Streamflow observation data for the selected river basins were retrieved from the Global Runoff Data Centre (GRDC) [23] to compare with the hydrological model simulated discharge.

2.2.3 Climate scenarios

We used the output of GCMs from the first phase of the ISI-MIP [37], which has five GCMs from the CMIP5 archive. GCMs from the CMIP5 archive [38]. In the framework of the ISIMIP, five GCMs simulations were bi-linearly downscaled spatially to a 0.5° × 0.5° grid for the period of 1950–2099 [39]. The five GCMs are GFDL-ESM2M, HadGEM2-ES, IPSL-CM5A-LR, MIROC-ESM-CHEM, and NorESM1-M. The GCMs were selected because of their wide application for modelling, predicting climate extremes, and investigating physical climate processes over West Africa [12, 40–42]. Again, the availability of ensemble member bias-corrected using an established approach to facilitate climate change impact assessments [43], guided our selection.

2.3 Validation

Station records from GRDC for the majority of the rivers in West Africa have missing values. Validation of PCR-GLOBWB for the selected river systems was limited to four gauging stations, where we could get a minimum of 5 hydrological years. Five performance evaluation metrics were used to evaluate discharge simulated at the selected gauging stations [44, 45]. These statistics are the Kling–Gupta efficiency (KGE); the Nash–Sutcliffe Efficiency criteria (NSE); the coefficient of determination (r2) per cent bias (PBIAS); and the ratio of RMSE-observations standard deviation (RSR). The performance ratings for quantitative statistics are presented in **Table 2**.

2.4 Method of analyses

For the analyses, we used discharge climatologies, i.e. long-term average monthly river discharge of each river. Projected daily discharge from five climate models was averaged to monthly climatologies and analysed for two periods, the reference (1971–2000) and the far-future period (2070–2099). Absolute changes between Reference and far future conditions was computed for each climate model historical (1971–2000).

Performance Ratings	r^2	NSE	RSR	PBIAS	KGE
Very good	$0.75 < r^2 \leq 1$	$0.75 < NSE \leq$	$0 \leq RSR \leq 0.5$	PBIAS < ±10	$0.9 \leq KGE \leq 1$
Good	$0.65 < r^2 \leq 0.75$	$0.65 < NSE \leq 0.75$	$0.5 < RSR \leq 0.6$	±10 ≤ PBIAS < ±15	$0.75 \leq KGE < 0.9$
Satisfactory	$0.5 < r^2 \leq 0.65$	$0.5 < NSE \leq 0.65$	$0.6 < RSR \leq 0.7$	±15 ≤ PBIAS < ±25	$0.5 \leq KGE < 0.75$
Unsatisfactory	$r^2 \leq 0.5$	$NSE \leq 0.5$	$RSR > 0.7$	PBIAS ≥ ±25	$KGE < 0.5$

Table 2.
Overview of monthly performance ratings [45].

2.5 Seasonality index (SI)

The seasonality of river discharge volumes shows the extent of variation in monthly discharge magnitude throughout the year. In this study, we used SI, developed by [46]. SI is the summation of the absolute changes of the monthly river discharge volumes from the mean monthly discharge, divided by the total annual river discharge of a given year. It is given as:

$$SI = \frac{\sum_{n=1}^{12} X_n - \frac{R}{12}}{R} \tag{1}$$

where X_n is the mean discharge of the nth month, and R is the mean annual discharge. SI varies from zero (all months having equal discharge distribution) to 1.83 (the discharge occurs in one month). A seasonal pattern in the discharge regime is established when SI is 0.6 and above. A region with high SI would be susceptible to drought because a high SI value in an area translates to the high variability of water resources and the shortage with respect to time. The classification of the degree of SI values is presented in **Table 3**.

2.6 Results

2.6.1 PCR-GLOBWB validation

PCR-GLOBWB models' validation results in terms of five metrics are presented in **Table 4** and **Figure 2** for the gauges. Overall, the performance of the PCR-GLOBWB on the monthly time step for validation periods was satisfactory. As seen in **Figure 2**, the hydrographs monthly flow pattern was well reproduced, close agreement in discharge distribution is seen in accord with performance statistics. The PCR-GLOBWB model performance was adequate in all basins, displaying very good to satisfactory values, established on ratings detailed in [44, 45]. **Table 5** sums up the five performance metrics values obtained for validation periods in each study basin. Concerning KGE values, the best model fit was found for the Hadejia (0.88) and Yobe (0.87) basins. The Niger, Hadejia, and Yobe river have very good NSE (0.8, 0.79, and 0.79), RSR (0.4, 0.45, and 0.4), and r^2 (0.88, 0.79, and 0.82) values, while Jamaare has satisfactory values (NSE = 0.6, RSR = 0.62, r^2 = 0.62). The highest PBIAS values reaching −18 to −25% were obtained in the Niger and Jamaare river. The PCR-GLOBWB model mostly reproduced well the flow dynamics of the

Seasonality Index	Regime
< 0.19	Very equable (discharge equally spread all over moths)
0.20–0.39	Equable, but with a definite wetter period
0.40–0.59	Rather seasonal with a short drier period
0.60–0.79	Seasonal
0.80–0.99	Marked seasonal with a long dry period
1.00–1.19	Most discharge in < 3 months
>1.20	Extreme discharge, with almost all discharges in 1–2 months

Table 3.
Classification of seasonality index (SI).

River	Validation period	KGE	NSE	R^2	RSR	PBIAS %
Niger	1979–1989	0.73	0.8	0.88	0.4	−25
Jama'are	1966–1975	0.69	0.6	0.62	0.62	−18
Yobe	1978–1983	0.87	0.79	0.82	0.4	6.68
Hadejia	1982–1989	0.88	0.79	0.79	0.45	0.8

Table 4.
Model's validation performance for the river basins.

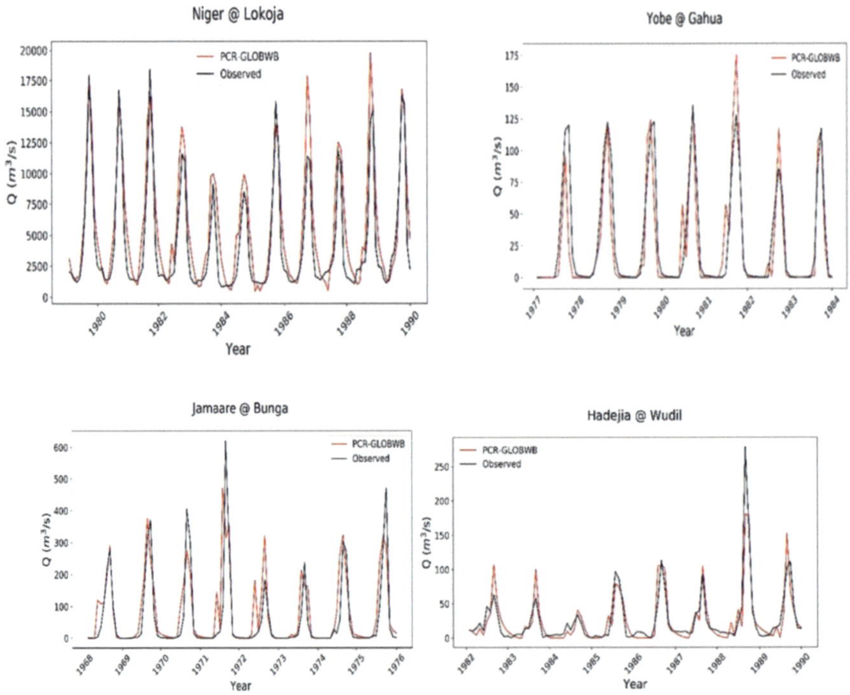

Figure 2.
Hydrographs for the different validation period of the four rivers (a) Niger (b) Yobe (c) Jamaare (d) Hadejia. Red line is PCR-GLOBWB and Black line is Observed.

	Reference-period	RCP4.5	RCP8.5
Niger	0.56 (0.51–0.59)	0.58 (0.48–0.62)	0.61 (0.45–0.63)
Yobe	1.05 (1.03–1.18)	1.09 (0.97–1.30)	1.03 (0.98–1.30)
Jamaare	1.07 (1.04–1.09)	1.14 (0.97–1.22)	1.12 (0.96–1.25)
Hadejia	0.89 (0.85–0.91)	0.86 (0.68–0.92)	0.87 (0.69–0.93)

Table 5.
Seasonality index (SI) of climate models median discharge climatologies for reference and projected far-future periods under RC4.6 and RCP8.5. Minimum and maximum values of the climate models' combinations are shown in brackets.

observed but displayed some disparity. Peak values overestimation was also seen in most cases for the river basins. The overestimation of flow can be attributed to the CRU TS 32.1, forcing input data. CRU Precipitation across Africa is of low quality due to sparse CRU stations and limited data assimilated during the reanalysis of ERA-40 [47]. As a result, of the temporal and spatial disparity in station density, the datasets are subjected to uncertainties; this uncertainty explains the overestimation of the stations' hydrographs. However, CRU TS 3.2 is a preferred data as it based on observation. Using another meteorological dataset may reduce the overestimation, but at the cost of the temporal variability, because no other datasets cover a long period of 1958–2015.

2.6.2 Historical discharge representation

Simulated discharge of the PCR-GLOBWB driven with five climate models was compared with CRU observed discharge to check the model's performance for the reference period (**Figure 3**). **Figure 3** shows the agreement between CRU observed discharge and the climate models is good for the Niger and Hadejia, yet there are some differences in representing high flow periods. For the Yobe and Jamaare rivers,

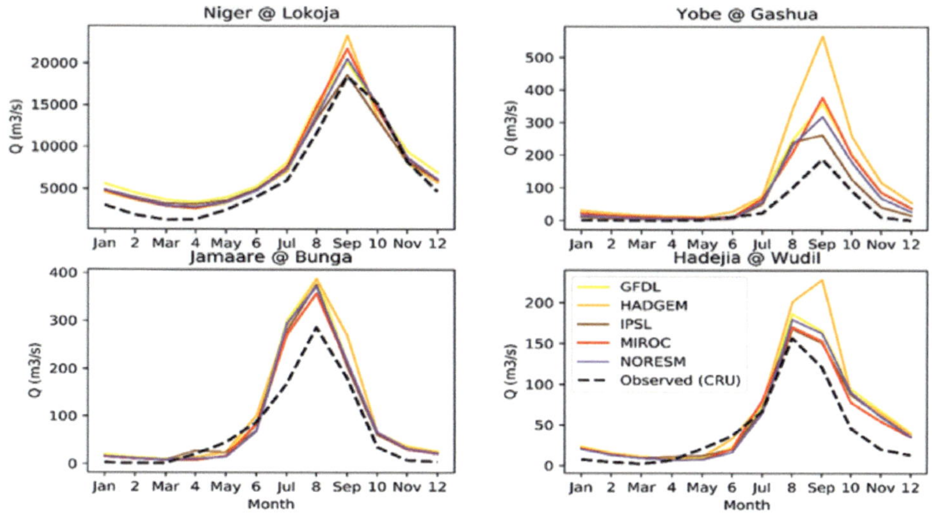

Figure 3.
The long-term mean annual dynamics of river discharge simulated with PCR-GLOBWB driven by the five climate models outputs over the reference period 1971–2000 for the four case study rivers.

the seasonal pattern of observed discharge dynamics was well captured; however, the results differ more distinctly in the representation of high flow periods. Nonetheless, these results are still agreeable, considering the small absolute numbers of discharge in these basins.

2.6.3 Evaluation of climate trends in the basins

Figure 4 presents the seasonal dynamics of precipitation and **Figure 5** of temperature for the reference and far future periods as illustrated by the median under RCP4.5 and RCP8.5 scenarios. Increasing temperature is projected in all study basins all year round under both scenarios, plus higher increases under the RCP8.5 compared to the RCP4.5. The highest increase in temperature is observed for the

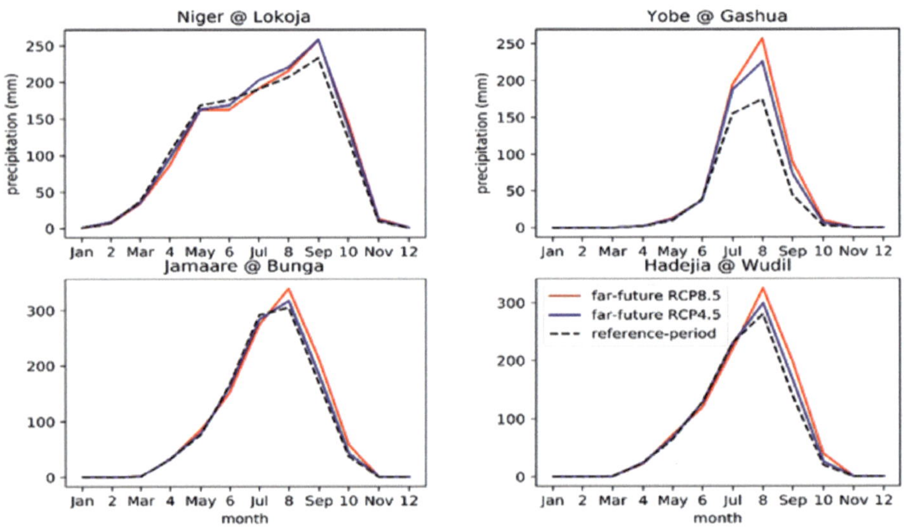

Figure 4.
Long-term monthly precipitation for the reference and far-future time slices as indicated by the climate models median of the climate projections under RCP 4.5 and RCP 8.5.

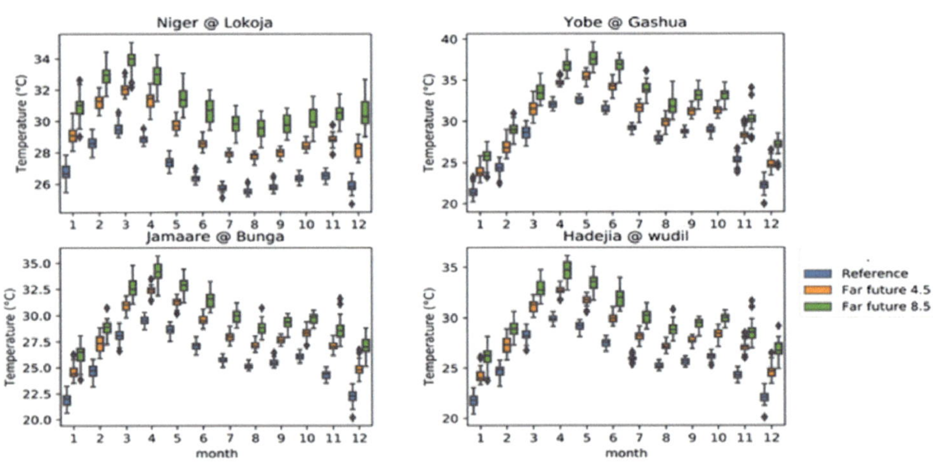

Figure 5.
Long-term monthly temperature for the reference and far-future time slices as indicated by the climate models median of the climate projections under RCP 4.5 and RCP 8.5.

Yobe River, reaching about 38°C in the month of may (RCP 8.5). Considering the precipitation dynamics in the Lokoja gauge of the Niger River located in central Nigeria, a moderate decrease at the beginning of the rainy season (from March to July), under both RCPs, is observed. Additionally, increasing precipitation is projected in Niger from July to November. A similar trend of a moderate reduction in rainfall at the start of the rainy season (June–July) is also observed for the Jama'are and Hadejia rivers. At the same time, an eventual increase is seen from July to November in the latter part of the year. The increase is from July to October in the Yobe, Jamaare, and Hadejia under both scenarios. Increasing precipitation is projected from June to November at Gashua (Yobe river). Biasutti [48] discovered decreasing rainfall in the start months (June–July) and increased rainfall in the end months (September–October), signifying a delay of the rainy season in West Africa. Refs. [49–51] also found a late beginning of the rainy season across Sahelian Africa because of a decrease in precipitation in July and August.

2.6.4 The seasonality index

As explained previously, we quantified climate change impacts on the monthly concentration and seasonal variation of discharge volumes. **Table 5** presents the SI of the five GCMs median considering the reference period climate and far-future climate under RCP4.5 and RCP8.5. As seen from the table, variations in SI between reference-period and end-century regimes are overall small. Over the Niger basin, the annual discharge volume under the RCP8.5 is seasonal compared with the reference period. For the Yobe and Jamaare rivers, the future annual river discharge volumes would be concentrated in less than three months. Annual SI over the HJKYRB river for the reference and Far-future period shows similar seasonal variability regimes of marked seasonality with a long dry period. The two basins generally receive most of the precipitation during the wet season; therefore, annual discharge occurs during the high-flow period.

2.6.5 Climate change impacts on seasonal discharge variation

Multi-model median monthly river discharge climatologies for the reference and far-future conditions under scenarios are presented in **Figure 6**.

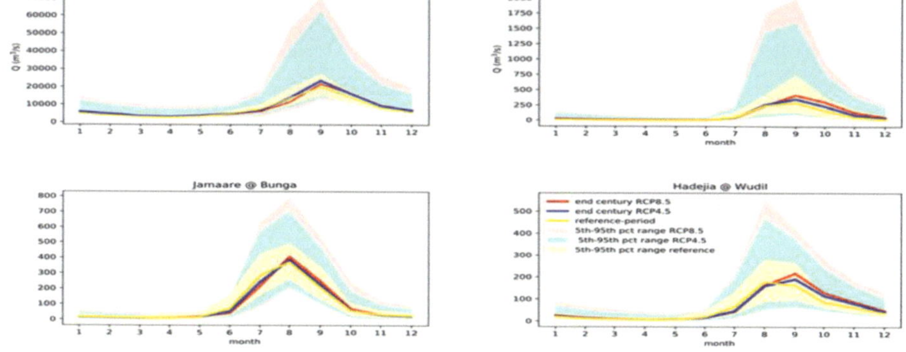

Figure 6.
Monthly streamflow climatologies projected for reference (1971–2000) and far-future (2070–2099) periods. The solid lines represent the ensemble median of all climate models combinations, while the colours represent the range from the 5th to 95th percentile.

2.6.5.1 Niger Basin at the Lokoja gauge station

The multi-model median reveals a reduction in the river discharge at the start of the high flow period. Total projected decreases added up to 512 m^3/s under RCP 4.5 (June to July) and 3652 m^3/s under RCP 8.5 from June to August. The most reduction in discharge volume (2101 m^3/s) occurs in August under RCP8.5, corresponding to about 58% of the total decrease for the three months. Eisner et al. [13] showed that under RCP8.5, the Niger basin shows declining discharge at the onset of the rainy season, with the highest loss 37 per cent occurring in August. When looking under the RCP 4.5 (**Figure 6**), discharge volume is expected to increase from August. The most significant increase under the moderate end scenario is shown in September and October, which considerably adds up to about +5565 m3/s (+52%) volume to the total increase in discharge volume. **Figures** 7 and **8** show that MIROC-ESM-CHEM hydrological simulation under both scenarios results in considerable volume increases all through the year caused by a related rise in precipitation, particularly towards the end of the rain period (August to September). **Figures** 7 and **8** also reveal that the other four hydrological simulations show contrasting change direction (decrease and increase) notably from August to November based on the individual climate model. The conflicting trend of change of each climate model cancels out in the multi-model median.

2.6.5.2 Hadejia-Jama'are Komadugu-Yobe Basin (HJKYB)

2.6.5.2.1 Yobe river at Gashua gauge station

Precipitation pattern influences river discharge seasonality in the Yobe river. Although the broad seasonal change signal at the Gashua gauge is unaltered by climate change, the climate models median show, considerable increases in river discharge volumes from August to November (**Figure 6**). Discharge volume increased by + 371 m^3/s under RCP8.5 and + 191 m^3/s under RCP4.5 for the four-month period, amounting to about +87% (RCP4.5) and +90% (RCP4.5) of the total increased volumes. Generally, this increase in discharge arises from two climate models (**Figures** 7 and **8**): the MIROC-ESM-CHEM and HADGEM-ES result in increased precipitation in the rainy season (July to October).

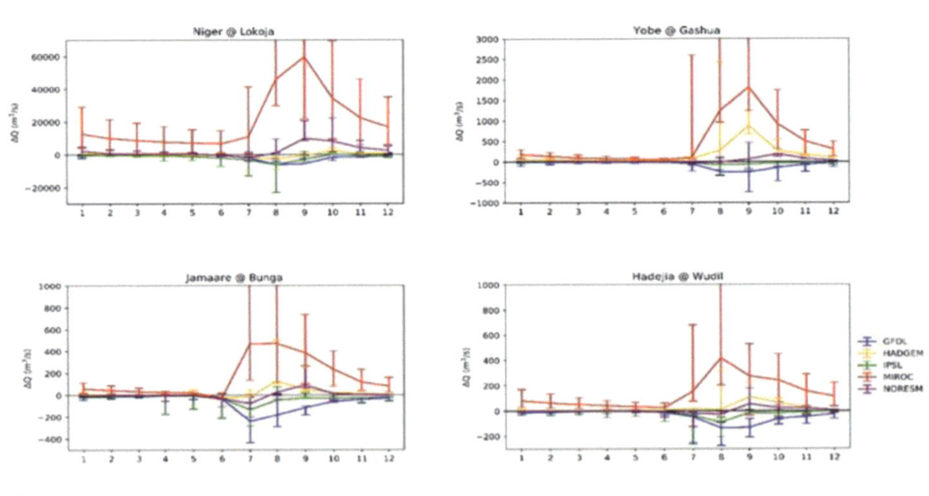

Figure 7.
Absolute change in discharge climatology projected under RCP8.5 for each GCM.

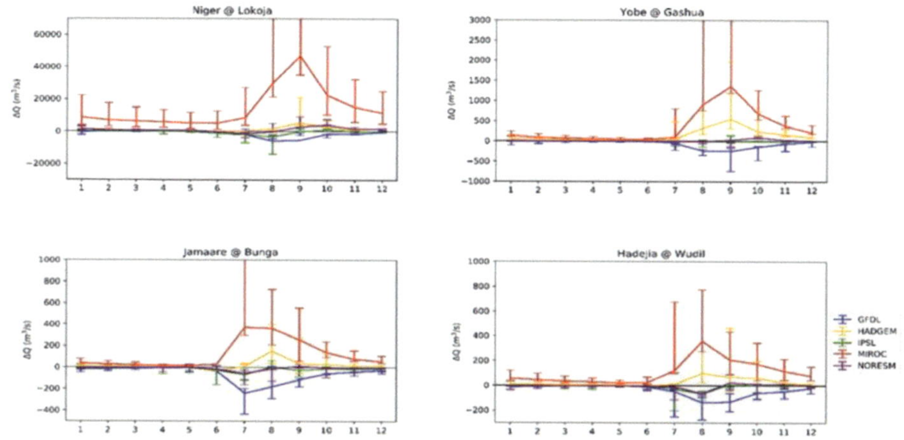

Figure 8.
Absolute change in discharge climatology projected under RCP4.5 for each GCM.

Hydrological projections of NORESM and IPSL-CM5A-LR show minor changes from the reference period, while the GFDL decreases the discharge volume (**Figures** 7 and **8**).

2.6.5.2.2 Jamaare river at the Bunga gauge station

The precipitation change pattern mainly influences the Jamaare river discharge regime. **Figure 6** reveals a one-month shift and decreases in discharge volumes from May to July at the start of the high flow season under both scenarios. The expected decrease in precipitation amounts explains this decrease in discharge volume (**Figure 4**). Under the RCP8.5, a reduction of 32 mm in precipitation for June to July translates to a reduction of -91 m^3/s (−24%). For the moderate end scenario, a decrease of 11 mm in precipitation results in decreasing discharge volume of -40 ms/s (−14%). An increase in discharge volume is shown from August to October in the latter part of the year. Projected increases in discharge volumes from August–October amount to +84 m^3/s (+13%) under RCP8.5 and +31 $m^3/3$ (+4%) for RCP4.5. In absolute terms, under both RCPs, (in **Figures** 7 and **8**), discharge response of the high flow season (June to October) is conflicting, having projected decreasing and increasing trend dependent on the GCM.

2.6.5.2.3 Hadejia river at Wudil gauge station

The hydrograph for the Hadejia river presents the peak monthly streamflow shifting from August to September for two scenarios (**Figure 6**). This shift is likely linked to a delayed start of precipitation. The climate model median projects a reduction in discharge volumes from June to August amounting to −39 m3/s (−15%) under RCP8.5 and −40 m3/s (−15%) under RCP 4.5. An increase in river discharge volumes is shown from September to December under both RCPs, which is related to increasing precipitation. For the RCP 8.5, the cumulative increase of discharge volume is from September to December amounts to 133 m3/s (38%) for RCP8.5 and 80 m3/s (23%) for RCP4.5. When considering the individual GCMs, Three out of the five climate models show increasing volumes in discharge in the peak flow period for RCP8.5 (**Figure 8**). The largest increase is found for MIROC (**Figures** 7 and **8**); however, the signal of change depends on the Individual model.

In all four rivers, hydrological simulations of the MIROC climate model reveal the highest discharge compared to the other four climate models in all cases.

2.7 Discussion

The evaluation of climate change's hydrological consequences often necessitates using a hydrological model forced with GCMs under various emission scenarios [52]. This study established the use of the PCR-GLOBWB model for two West African river basins; the Niger river basin and the Hadejia-Jama'are Komadugu-Yobe Basin (HJKYB). The monthly streamflow simulation results of the gauge stations of the two river basins were validated against observed discharge (GRDC) using five performance evaluation metrics were. The output of GCMs from the first phase of the ISI-MIP was adopted to investigate the response of streamflow seasonality to climate change. The PCR-GLOBWB was shown to be very applicable over the two basins. Its PBIAS, NSE, RSR, r^2 and KGE values ranged from −25 to 0.8, 0.6 to 0.8, 0.62 to 0.4, 0.62–0.88, and 0.69 to 0.88, respectively, which were within the acceptable limits [45], as shown in **Table 2**.

According to the results of the multi-model median regarding climate change, climate change impacted the temporal pattern of future river discharge in the river basins. The late start of the rainy season concluded in this work has been reported by previous studies [13, 48–50, 53]. Streamflow of the three rivers in combined HJKYB, the Yobe, Jamaare and the Hadejia, is controlled by precipitation. Genthon et al. [54] reported the climatic influence on discharge in this basin. Across the two basins, our findings indicate that climate change exacerbates the seasonality pattern already present. The basins influenced by precipitation exhibit a continuous increase in streamflow volumes during the later part of the high-flow season. In the Niger basin, climate change significantly affects the volume of streamflow seasonality (indicated by SI). In the four rivers, projected river discharge seasonality is amplified under the high-end emission scenario (RCP8.5); our findings support decisions on the potential advantages of reduced greenhouse gas emissions for the streamflow dynamics. It must be stressed that our analysis focused exclusively on the effects of climate change on streamflow regimes. Population expansion and economic development envisaged for the future are expected to raise human demand for water resources, potentially intensifying their interference with the streamflow regime. The consequences of these could outweigh the climatic changes examined in this study.

3. Conclusions

This study assessed of climate change impacts on the seasonal river discharge in two rivers in West Africa, including the Niger, and the Hadejia-Jama'are Komadugu-Yobe Basin (HJKYB). For this analysis, we set up and validated the PCR-GLOBWB model at the selected gauging stations of each river basin. Climate change impacts on river discharge seasonality were examined with five bias-corrected GCMs, collected from the "ISIMIP" project framework. The PCR-GLOBWB model validation performance was satisfactory in its performance for all statistics at each of the basins (**Table 5**). The five bias-corrected GCMs were then used to force PCR-GLOBWB in the reference and far-future period. Based on our results, Climate change will influence the seasonal regime of discharge of the selected rivers, i.e., the timing and the magnitude of flows. The findings of this study reveal that there are little differences in SI between the present and the far-future. However, climate change will affect the temporal seasonality pattern. At the gauges of the Niger, Yobe, and Jamaare rivers, decreasing discharge volumes when the high-flow period begins (typically May–July) is expected in the far future. This is explained by the delayed start of the raining season. All four rivers steadily project increasing river discharge at the end of the high-flow season (typically August–November) during the peak-flow period. In this basins, increased precipitation amounts result in a projected increase in discharge volumes.

Adequate storage has to be made for the increased high-flow season; otherwise, water scarcity may disturb agricultural production regardless of the overall increases in annual water availability. Even though increased discharge volumes can be considered advantageous to agricultural productivity, this can only be achievable when there are provisions to store excess flow for later use. When there are no/sufficient existent storage structures, much of the additional flow is lost. Increased high flow may destroy croplands through flooding. Increased discharge volumes could lead to floods and destroy crops over a vast expanse of land. Excess water could reduce plant development, delay farm operations, make the soil soggy and unworkable. The projected increased discharge for the HJRB could help revive the currently shrinking Lake Chad basin. The findings of this study show that climate change will significantly impact the hydrological regimes of the two basins examined, with significant consequences for water resource planning and management. Finally, the methods used in this study may prove helpful for future research examining the effects of climate change on the hydrology of different regions.

Acknowledgements

We appreciate Utrecht University's Department of Physical Geography, notably Edwin Sutanudjaja and Rens van Beek, for providing the PCR-GLOBWB model and input datasets.

Conflict of interest

The authors declare no conflict of interest.

Author details

Toju Esther Babalola[1*], Philip Gbenro Oguntunde[2], Ayodele Ebenezer Ajayi[2] and Francis Omowonuola Akinluyi[2]

1 Department of Water Resources Management and Agro-Meteorology, Federal University Oye-Ekiti, Ekiti State, Nigeria

2 Department of Agricultural Engineering, Federal University of Technology Akure, Ondo State, Nigeria

*Address all correspondence to: toju.babalola@fuoye.edu.ng

References

[1] Chu JT, Xia J, Xu CY, Singh VP. Statistical downscaling of daily mean temperature, pan evaporation and precipitation for climate change scenarios in Haihe River, China. Theor Appl Climatol. 2010;99(1-2).

[2] Chien H, Yeh PJF, Knouft JH. Modeling the potential impacts of climate change on streamflow in agricultural watersheds of the Midwestern United States. J Hydrol. 2013;491(1).

[3] Bodian A, Dezetter A, Diop L, Deme A, Djaman K, Diop A. Future climate change impacts on streamflows of Two Main West Africa River Basins: Senegal and Gambia. Hydrology. 2018;

[4] Sonwa DJ, Dieye A, El Mzouri EH, Majule A, Mugabe FT, Omolo N, et al. Drivers of climate risk in African agriculture. Clim Dev. 2017;

[5] Rebelo LM, McCartney MP, Finlayson CM. Wetlands of Sub-Saharan Africa: Distribution and contribution of agriculture to livelihoods. Wetl Ecol Manag. 2010;

[6] Ola O, Benjamin E. Preserving biodiversity and ecosystem services in West African forest, watersheds, and wetlands: A review of incentives. Forests. 2019;

[7] Lee D, Ward P, Block P. Defining high-flow seasons using temporal streamflow patterns from a global model. Hydrol Earth Syst Sci. 2015;

[8] Wang HM, Chen J, Xu CY, Chen H, Guo S, Xie P, et al. Does the weighting of climate simulations result in a better quantification of hydrological impacts? Hydrol Earth Syst Sci. 2019;23(10).

[9] Lv X, Zuo Z, Ni Y, Sun J, Wang H. The effects of climate and catchment characteristic change on streamflow in a typical tributary of the Yellow River. Sci Rep. 2019;9(1).

[10] Teutschbein C, Seibert J. Bias correction of regional climate model simulations for hydrological climate-change impact studies: Review and evaluation of different methods. J Hydrol. 2012;456-457.

[11] Dettinger MD, Diaz HF. Global characteristics of stream flow seasonality and variability. J Hydrometeorol. 2000;

[12] Aich V, Liersch S, Vetter T, Huang S, Tecklenburg J, Hoffmann P, et al. Comparing impacts of climate change on streamflow in four large African river basins. Hydrol Earth Syst Sci. 2014;

[13] Eisner S, Flörke M, Chamorro A, Daggupati P, Donnelly C, Huang J, et al. An ensemble analysis of climate change impacts on streamflow seasonality across 11 large river basins. Clim Change. 2017;

[14] Hirpa FA, Alfieri L, Lees T, Peng J, Dyer E, Dadson SJ. Streamflow response to climate change in the Greater Horn of Africa. Clim Change. 2019;

[15] WaleWorqlul A, Taddele YD, Ayana EK, Jeong J, Adem AA, Gerik T. Impact of climate change on streamflow hydrology in headwater catchments of the upper Blue Nile Basin, Ethiopia. Water (Switzerland). 2018;

[16] Li Z, Jin J. Evaluating climate change impacts on streamflow variability based on a multisite multivariate GCM downscaling method in the Jing River of China. Hydrol Earth Syst Sci. 2017;

[17] Ficklin DL, Stewart IT, Maurer EP. Climate change impacts on streamflow and subbasin-scale hydrology in the Upper Colorado River Basin. PLoS One. 2013;

[18] Vano JA, Nijssen B, Lettenmaier DP. Seasonal hydrologic responses to climate change in the P acific N orthwest. Water Resour Res [Internet]. 2015 Jun 25;51(4):1959-76. Available from: https://onlinelibrary.wiley.com/doi/abs/10.1002/2014WR015909

[19] Boko M, Niang I, Nyong A, Vogel C, Githeko A, Medany M, et al. Africa. Climate Change 2007: Impact, Adaptation and Vulnerabilitiy. In: Contribution of Working Group II to the Fourth Assessment Report of the Intergovernmental Panel on Climate Change. 2007.

[20] Ayeni AO, Kapangaziwiri E, Soneye ASO, Engelbrecht FA. Assessing the impact of global changes on the surface water resources of Southwestern Nigeria. Hydrol Sci J [Internet]. 2014 Jul 15;150527103244004. Available from: http://www.tandfonline.com/doi/full/10.1080/02626667.2014.993645

[21] Ndulue EL, Mbajiorgu CC. Modeling climate and landuse change impacts on streamflow and sediment yield of an agricultural watershed using SWAT. Agric Eng Int CIGR J. 2018;

[22] Sylla MB, Faye A, Klutse NAB, Dimobe K. Projected increased risk of water deficit over major West African river basins under future climates. Clim Change. 2018;

[23] Bierkens MFP. Global hydrology 2015: State, trends, and directions. Water Resour Res. 2015;

[24] Beck HE, Zimmermann NE, McVicar TR, Vergopolan N, Berg A, Wood EF. Present and future köppen-geiger climate classification maps at 1-km resolution. Sci Data. 2018;

[25] Femi Monday I. Investigating Effects of Climate Change on Health Risks in Nigeria. In: Environmental Factors Affecting Human Health [Working Title] [Internet]. IntechOpen; 2019. Available from: https://www.intechopen.com/online-first/investigating-effects-of-climate-change-on-health-risks-in-nigeria

[26] Elisha I, B. A S, Lawrence EU. Evidence of Climate Change and Adaptation Strategies among Grain Farmers in Sokoto State, Nigeria. IOSR J Environ Sci Toxicol Food Technol [Internet]. 2017 Jul 14;11(03):1-7. Available from: http://www.iosrjournals.org/iosr-jestft/papers/vol11-issue 3/Version-2/A1103020107.pdf

[27] Aich V, Koné B, Hattermann FF, Paton EN. Time series analysis of floods across the Niger River Basin. Water (Switzerland). 2016;

[28] Ogilvie A, Mahé G, Ward J, Serpantié G, Lemoalle J, Morand P, et al. Water, agriculture and poverty in the niger river basin. Water Int. 2010;

[29] Ejieji C, Amodu M, Adeogun A. Prediction of the streamflow of Hadejia-Jama'are-Komadugu-Yobe-River Basin, North Eastern Nigeria, using swat model. Ethiop J Environ Stud Manag. 2016;

[30] Umar AS, Ankidawa BA. Climate Variability and Basin Management : A Threat to and from Wetlands of Komadugu Yobe Basin , North Eastern Nigeria. 2016;04(02):25-36.

[31] Sutanudjaja EH, Van Beek R, Wanders N, Wada Y, Bosmans JHC, Drost N, et al. PCR-GLOBWB 2: A 5 arcmin global hydrological and water resources model. Geosci Model Dev. 2018;

[32] Wada Y, de Graaf IEM, van Beek LPH. High-resolution modeling of human and climate impacts on global water resources. J Adv Model Earth Syst. 2016;

[33] Karssenberg D, de Jong K, van der Kwast J. Modelling landscape dynamics with Python. Int J Geogr Inf Sci. 2007;

[34] Harris I, Jones PD, Osborn TJ, Lister DH. Updated high-resolution grids of monthly climatic observations - the CRU TS3.10 Dataset. Int J Climatol. 2014;

[35] Uppala SM, Kållberg PW, Simmons AJ, Andrae U, da Costa Bechtold V, Fiorino M, et al. The ERA-40 re-analysis. Quarterly Journal of the Royal Meteorological Society. 2005.

[36] Dee DP, Uppala SM, Simmons AJ, Berrisford P, Poli P, Kobayashi S, et al. The ERA-Interim reanalysis: Configuration and performance of the data assimilation system. Q J R Meteorol Soc. 2011;

[37] Asadieh B, Krakauer NY. Global change in streamflow extremes under climate change over the 21st century. Hydrol Earth Syst Sci. 2017;

[38] Agyekum J, Annor T, Lamptey B, Quansah E, Agyeman RYK. Evaluation of CMIP5 Global Climate Models over the Volta Basin: Precipitation. Adv Meteorol. 2018;

[39] Stagl JC, Hattermann FF. Impacts of climate change on the hydrological regime of the danube river and its tributaries using an ensemble of climate scenarios. Water (Switzerland). 2015;

[40] Akumaga U, Tarhule A, Piani C, Traore B, Yusuf AA. Utilising process-based modeling to assess the impact of climate change on crop yields and adaptation options in the Niger river Basin, West Africa. Agronomy. 2018;

[41] Oyerinde G, Hountondji F, Lawin A, Odofin A, Afouda A, Diekkrüger B. Improving Hydro-Climatic Projections with Bias-Correction in Sahelian Niger Basin, West Africa. Climate [Internet]. 2017 May 11;5(1):8. Available from: http://www.mdpi.com/2225-1154/5/1/8

[42] Thompson JR, Crawley A, Kingston DG. Future river flows and flood extent in the Upper Niger and Inner Niger Delta: GCM-related uncertainty using the CMIP5 ensemble. Hydrol Sci J. 2017;

[43] Gosling SN, Zaherpour J, Mount NJ, Hattermann FF, Dankers R, Arheimer B, et al. A comparison of changes in river runoff from multiple global and catchment-scale hydrological models under global warming scenarios of 1 °C, 2 °C and 3 °C. Clim Change. 2017;

[44] D. N. Moriasi, J. G. Arnold, M. W. Van Liew, R. L. Bingner, R. D. Harmel, T. L. Veith. Model Evaluation Guidelines for Systematic Quantification of Accuracy in Watershed Simulations. Trans ASABE. 2007;

[45] Kouchi DH, Esmaili K, Faridhosseini A, Sanaeinejad SH, Khalili D, Abbaspour KC. Sensitivity of calibrated parameters and water resource estimates on different objective functions and optimisation algorithms. Water (Switzerland). 2017;

[46] Walsh RPD, Lawler DM. RAINFALL SEASONALITY: DESCRIPTION, SPATIAL PATTERNS AND CHANGE THROUGH TIME. Weather. 1981;

[47] Van Beek LPH, Wada Y, Bierkens MFP. Global monthly water stress: 1. Water balance and water availability. Water Resour Res. 2011;

[48] Biasutti M. Forced Sahel rainfall trends in the CMIP5 archive. J Geophys Res Atmos. 2013;118(4).

[49] Dunning CM, Black E, Allan RP. Later wet seasons with more intense rainfall over Africa under future climate change. J Clim. 2018;31(23).

[50] Patricola CM, Cook KH. Northern African climate at the end of the twenty-first century: An integrated

application of regional and global climate models. Clim Dyn. 2010;

[51] Wang G, Alo CA. Changes in precipitation seasonality in West Africa predicted by RegCM3 and the impact of dynamic vegetation feedback. Int J Geophys. 2012;

[52] Guo Y, Fang G, Xu YP, Tian X, Xie J. Identifying how future climate and land use/cover changes impact streamflow in Xinanjiang Basin, East China. Sci Total Environ. 2020;710.

[53] Biasutti M, Sobel AH. Delayed Sahel rainfall and global seasonal cycle in a warmer climate. Geophys Res Lett. 2009;

[54] Genthon P, Hector B, Luxereau A, Descloitres M, Abdou H, Hinderer J, et al. Groundwater recharge by Sahelian rivers—consequences for agricultural development: example from the lower Komadugu Yobe River (Eastern Niger, Lake Chad Basin). Environ Earth Sci. 2015;74(2).